Lab Workbook

# Welding
## FUNDAMENTALS

by

**William A. Bowditch**
Career and Technical Education Consultant, Portage, Michigan
Life Member of the American Welding Society
Member of the Association for Career and Technical Education

**Kevin E. Bowditch**
Welding Engineer Specialist, Subaru of Indiana Automotive, Inc., Lafayette, Indiana
Member of the American Welding Society
Member of the Association for Career and Technical Education

**Mark A. Bowditch**
Member of the American Welding Society
Member of the Association for Career and Technical Education

Publisher
**The Goodheart-Willcox Company, Inc.**
Tinley Park, IL
www.g-w.com

**The Goodheart-Willcox Company, Inc. Brand Disclaimer:** Brand names, company names, and illustrations for products and services included in this text are provided for educational purposes only and do not represent or imply endorsement or recommendation by the author or the publisher.

**The Goodheart-Willcox Company, Inc. Safety Notice:** The reader is expressly advised to carefully read, understand, and apply all safety precautions and warnings described in this book or that might also be indicated in undertaking the activities and exercises described herein to minimize risk of personal injury or injury to others. Common sense and good judgment should also be exercised and applied to help avoid all potential hazards. The reader should always refer to the appropriate manufacturer's technical information, directions, and recommendations; then proceed with care to follow specific equipment operating instructions. The reader should understand these notices and cautions are not exhaustive.

The publisher makes no warranty or representation whatsoever, either expressed or implied, including but not limited to equipment, procedures, and applications described or referred to herein, their quality, performance, merchantability, or fitness for a particular purpose. The publisher assumes no responsibility for any changes, errors, or omissions in this book. The publisher specifically disclaims any liability whatsoever, including any direct, indirect, incidental, consequential, special, or exemplary damages resulting, in whole or in part, from the reader's use or reliance upon the information, instructions, procedures, warnings, cautions, applications, or other matter contained in this book. The publisher assumes no responsibility for the activities of the reader.

**The Goodheart-Willcox Company, Inc. Internet Disclaimer:** The Internet resources and listings in this Goodheart-Willcox Publisher product are provided solely as a convenience to you. These resources and listings were reviewed at the time of publication to provide you with accurate, safe, and appropriate information. Goodheart-Willcox Publisher has no control over the referenced websites and, due to the dynamic nature of the Internet, is not responsible or liable for the content, products, or performance of links to other websites or resources. Goodheart-Willcox Publisher makes no representation, either expressed or implied, regarding the content of these websites, and such references do not constitute an endorsement or recommendation of the information or content presented. It is your responsibility to take all protective measures to guard against inappropriate content, viruses, or other destructive elements.

# Introduction

This Lab Workbook is intended for use with the textbook, *Welding Fundamentals.* It will help you more thoroughly explore the concepts presented in the textbook.

The *Welding Fundamentals* textbook and this Lab Workbook are divided into eight major sections. Each section deals with a different welding process or area related to the welding field. Section 1, Introduction to Welding, explores safety in the welding shop, provides an introduction to welding careers, defines the various welding and cutting processes, explains the physics of welding, introduces important welding-related math concepts and applications, describes the various weld joints and welding positions, and explains welding symbols and their function in drawings.

In Sections 2 through 7, extensive instruction is provided for various welding or cutting processes. The topics covered in each of these Sections includes equipment, equipment assembly and adjustment, cutting and/or welding techniques for all positions, and, in some cases, discussion of related processes. Shielded metal arc welding (SMAW) and surfacing are explained in Section 2. Section 3 deals with the gas metal arc welding (GMAW) and flux cored arc welding (FCAW). The gas tungsten arc welding (GTAW) process is the focus of Section 4. Section 5 offers an in-depth look at the plasma arc cutting (PAC) process. Section 6 contains detailed coverage of oxyfuel gas processes, including oxyfuel gas welding (OFW), oxyfuel gas cutting (OFC), braze welding (BW), brazing (B), and soldering (S). Various resistance welding (RW) processes are discussed in Section 7.

The final section provides information to help you expand your horizons as a welder. Section 8 includes helpful information about welding pipe and tube, information about robotics in welding, a survey of some special cutting and welding processes, coverage of inspecting and testing welds, and information about welder certification.

You may start with Section 1, Introduction to Welding, and progress in sequence through Section 8. As an alternative, you may study the Sections in any order. Regardless of the order you choose, it is highly recommended that you first study Section 1, Introduction to Welding, including Chapter 8, Welding Symbols. Becoming familiar with the AWS welding symbol is important because it is used in the later lessons of this Lab Workbook to provide vital information about the assigned welding jobs. Sizes on the welding symbols are provided in inches and fractions of an inch (or decimals). Equivalent measurements in the SI metric system are provided as well. You should note that the drawings in this Lab Workbook are not to scale.

Each Lesson in this Lab Workbook contains objectives and instructions. Several types of questions—identification, true or false, short completion, longer written answers, and multiple choice—are asked in each Lesson. Do your best to answer these questions carefully and accurately.

Sections 2 through 7 begin with a Safety Test that tests your knowledge of the safety considerations related to the welding and/or cutting processes covered in that section. You must pass the Safety Test and receive your instructor's permission before performing the Assigned Jobs for that section.

Before performing an Assigned Job, read all directions carefully, paying special attention to the safety precautions outlined in the *Welding Fundamentals* text and in this Lab Workbook. Inspection criteria and test procedures will be given for each Assigned Job. You should inspect your own work before submitting it to your instructor for grading.

You may wish to keep the Lessons, Assigned Jobs, and Safety Tests after they are graded. These pages can be shown to a prospective employer to document the extent and quality of your welding experiences.

# Table of Contents

## Section 1
## Introduction to Welding

**Lesson 1**—Careers in Welding . . . . . . . . . . . 7
**Lesson 2**—Safety in the Welding Shop . . . . . . 9
**Lesson 3**—Welding and Cutting Processes. . 11
**Lesson 4**—The Physics of Welding . . . . . . . 13
**Lesson 5**—Math for Welding . . . . . . . . . . . 15
**Lesson 6**—Math Applications for Welders . . . 19
**Lesson 7**—Weld Joints and Positions . . . . . . 23
**Lesson 8**—Welding Symbols. . . . . . . . . . . . 27
    Assigned Job 8-1—Reading an AWS
       Welding Symbol. . . . . . . . . . . . . 35
    Assigned Job 8-2—Drawing an AWS
       Welding Symbol. . . . . . . . . . . . . 39

## Section 2
## Shielded Metal Arc Welding

Shielded Metal Arc Welding Safety Test . . . . . 41
**Lesson 9**—SMAW: Equipment and Supplies . . . 45
**Lesson 10**—SMAW: Equipment Assembly and
    Adjustment . . . . . . . . . . . . . . . . . . . . . . . 49
    Assigned Job 10-1—Performing a Safety
       Inspection of a Shielded Metal Arc
       Welding Station . . . . . . . . . . . . . . 51
**Lesson 11**—SMAW: Electrodes . . . . . . . . . . 53
**Lesson 12**—SMAW: Flat Welding Position. . . 57
    Assigned Job 12-1—Running a Weld Bead
       in the Flat Welding Position . . . . . . 61
    Assigned Job 12-2—Welding a Lap Joint in
       the Flat Welding Position . . . . . . . . 63
    Assigned Job 12-3—Welding a T-Joint in
       the Flat Welding Position . . . . . . . . 65
    Assigned Job 12-4—Welding a V-Groove
       Outside Corner Joint in the Flat
       Welding Position . . . . . . . . . . . . . 67
**Lesson 13**—SMAW: Horizontal, Vertical, and
    Overhead Welding Positions. . . . . . . . . . . 69
    Assigned Job 13-1—Welding a Lap Joint in
       the Horizontal Welding Position . . . 73
    Assigned Job 13-2—Welding a T-Joint in
       the Horizontal Welding Position . . . 75
    Assigned Job 13-3—Welding a V-Groove
       Butt Joint in the Horizontal Welding
       Position . . . . . . . . . . . . . . . . . . . 77

    Assigned Job 13-4—Welding a Lap Joint in
       the Vertical Position . . . . . . . . . . . 79
    Assigned Job 13-5—Welding a T-Joint in
       the Vertical Welding Position . . . . . 81
    Assigned Job 13-6—Welding a V-Groove
       Butt Joint in the Vertical Welding
       Position . . . . . . . . . . . . . . . . . . . 83
    Assigned Job 13-7—Weldig a Lap Joint in
       the Overhead Welding Position . . . 85
    Assigned Job 13-8—Welding an Inside
       Corner Joint in the Overhead Welding
       Position . . . . . . . . . . . . . . . . . . . 87
    Assigned Job 13-9—Welding a V-Groove
       Butt Joint and an Outside Corner Joint
       in the Overhead Welding Position. . . 89
**Lesson 14**—Surfacing . . . . . . . . . . . . . . . . 91
    Assigned Job 14-1—Hardfacing Mild Steel
       with Hardfacing Electrodes and SMAW
       Equipment . . . . . . . . . . . . . . . . . 93

## Section 3
## Gas Metal and Flux Cored Arc Welding

Gas Metal and Flux Cored Arc Welding Safety
    Test. . . . . . . . . . . . . . . . . . . . . . . . . . . . 95
**Lesson 15**—GMAW and FCAW: Equipment and
    Supplies . . . . . . . . . . . . . . . . . . . . . . . . 97
**Lesson 16**—GMAW and FCAW: Equipment
    Assembly and Adjustment . . . . . . . . . . . .101
    Assigned Job 16-1—Learning to Use the
       Wire Feeder and Welding Gun. . . .103
    Assigned Job 16-2—Connecting the
       Shielding Gas Supply and Adjusting
       Gas Flow . . . . . . . . . . . . . . . . . .105
**Lesson 17**—GMAW and FCAW: Flat Welding
    Position. . . . . . . . . . . . . . . . . . . . . . . . 107
    Assigned Job 17-1—Making Stringer and
       Weave Beads. . . . . . . . . . . . . . . .109
    Assigned Job 17-2—Making Stringer
       Beads Using Spray Transfer. . . . . .111
    Assigned Job 17-3—Welding a Lap Joint in
       the Flat Welding Position . . . . . . . .113
    Assigned Job 17-4—Welding a T-Joint in
       the Flat Welding Position . . . . . . . .115
    Assigned Job 17-5—Welding a Square-
       Groove Butt Joint in the Flat Welding
       Position . . . . . . . . . . . . . . . . . . .117

**Lesson 18**—GMAW and FCAW: Horizontal, Vertical, and Overhead Welding Positions . .119
    Assigned Job 18-1—Welding a Lap Joint in the Horizontal Welding Position . . .121
    Assigned Job 18-2—Welding a T-Joint in the Horizontal Welding Position . . .123
    Assigned Job 18-3—Welding a Square-Groove Butt Joint in the Horizontal Welding Position . . . . . . . . . . . .125
    Assigned Job 18-4—Welding a Lap Joint in the Vertical Welding Position . . . . .127
    Assigned Job 18-5—Welding a T-Joint in the Vertical Welding Position . . . . .129
    Assigned Job 18-6—Welding a Square-Groove Butt Joint in the Vertical Welding Position . . . . . . . . . . . .131
    Assigned Job 18-7—Welding a Lap Joint in the Overhead Welding Position . . .133
    Assigned Job 18-8—Welding an Inside Corner Joint in the Overhead Welding Position . . . . . . . . . . . . . . . . . . .135
    Assigned Job 18-9—Welding a Square-Groove and V-Groove Butt Joint in the Overhead Welding Position . . . . . .137

**Section 4**
# Gas Tungsten Arc Welding

Gas Tungsten Arc Welding Safety Test . . . . . 139

**Lesson 19**—GTAW: Equipment and Supplies . 141

**Lesson 20**—GTAW: Equipment Assembly and Adjustment . . . . . . . . . . . . . . . . . . . . . . 145
    Assigned Job 20-1—Selecting and Preparing an Electrode . . . . . . . . .149
    Assigned Job 20-2—Assembling the Torch. .151

**Lesson 21**—GTAW: Flat Welding Position . . 153
    Assigned Job 21-1—Creating a Continuous Weld Pool. . . . . . . . . . . . . . . . . . .157
    Assigned Job 21-2—Welding a Lap Joint without Filler Metal. . . . . . . . . . . . .159
    Assigned Job 21-3—Welding a Square-Groove Butt Joint on an Outside Corner Joint without Filler Metal . .161
    Assigned Job 21-4—Making a Weave Bead on Plate . . . . . . . . . . . . . . . . . . . .163
    Assigned Job 21-5—Welding a Square-Groove Edge Joint without Filler Metal . . . . . .165
    Assigned Job 21-6—Making Stringer and Weave Beads on Plate with Filler Metal . . . . . . . . . . . . . . . . . . . . . .167

    Assigned Job 21-7—Making Stringer and Weave Beads on Stainless Steel with Filler Metal . . . . . . . . . . . . . . . . . .171
    Assigned Job 21-8—Making Stringer and Weave Beads on Aluminum with Filler Metal . . . . . . . . . . . . . . . . . . . . . .173
    Assigned Job 21-9—Welding a Lap Joint with Filler Metal in the Flat Welding Position . . . . . . . . . . . . . . . . . . . . .175
    Assigned Job 21-10—Welding a T-Joint with Filler Metal in the Flat Welding Position . . . . . . . . . . . . . . . . . . . . .177
    Assigned Job 21-11—Welding a Square-Groove Butt Joint in the Flat Welding Position . . . . . . . . . . . . . . . . . . . . .181

**Lesson 22**—GTAW: Horizontal, Vertical, and Overhead Welding Position. . . . . . . . . . . 183
    Assigned Job 22-1—Welding a Lap Joint in the Horizontal Welding Position . . .185
    Assigned Job 22-2—Welding a T-Joint in the Horizontal Welding Position . . .187
    Assigned Job 22-3—Welding a Square-Groove Butt Joint in the Horizontal Welding Position . . . . . . . . . . . . . .189
    Assigned Job 22-4—Welding a Lap Joint in the Vertical Welding Position . . . . .191
    Assigned Job 22-5—Welding a T-Joint in the Vertical Welding Position . . . . .193
    Assigned Job 22-6—Welding a Square-Groove Butt Joint in the Vertical Welding Position . . . . . . . . . . . . . .195
    Assigned Job 22-7—Welding a Lap Joint in the Overhead Welding Position . . .197
    Assigned Job 22-8—Welding an Inside Corner Joint in the Overhead Welding Position . . . . . . . . . . . . . . . . . . .199
    Assigned Job 22-9—Welding Square-Groove and V-Groove Butt Joints in the Overhead Welding Position . . . . . .201

**Section 5**
# Plasma Arc Cutting

Plasma Arc Cutting Safety Test. . . . . . . . . . 203

**Lesson 23**—Plasma Arc Cutting . . . . . . . . 205
    Assigned Job 23-1—Assembling Plasma Arc Cutting Equipment . . . . . . . . 207
    Assigned Job 23-2—Plasma Arc Cutting Mild Steel. . . . . . . . . . . . . . . . . . . 209
    Assigned Job 23-3—Plasma Arc Piercing and Cutting Mild Steel . . . . . . . . . .211

## Section 6
# Oxyfuel Gas Processes

Oxyfuel Gas Cutting and Welding Safety Test . . 213

**Lesson 24**—Oxyfuel Gas Cutting and Welding: Equipment and Supplies . . . . . . . . . . . . 217

**Lesson 25**—Oxyfuel Gas Cutting and Welding: Equipment Assembly and Adjustment . . 219
    Assigned Job 25-1—Turning On, Lighting, and Shutting Down the Oxyacetylene Outfit . . . . . . . . . . . . . . . . . . . . . . 221

**Lesson 26**—Oxyfuel Gas Cutting . . . . . . . 223
    Assigned Job 26-1—Manually Cutting 1/4″– 5/8″ (6.4 mm–15.9 mm) Mild Steel . . 227
    Assigned Job 26-2—Manually Cutting Mild Steel Less than 1/8″ (3.2 mm) Thick . 229
    Assigned Job 26-3—Cutting Mild Steel Using a Motorized Carriage and Track . . . . . 231

**Lesson 27**—Oxyfuel Gas Welding: Flat Welding Position . . . . . . . . . . . . . . . . . . . . . . . . . 233
    Assigned Job 27-1—Creating a Continuous Weld Pool on Mild Steel . . . . . . . . 237
    Assigned Job 27-2—Welding without a Welding Rod . . . . . . . . . . . . . . . . 239
    Assigned Job 27-3—Making Stringer and Weave Beads . . . . . . . . . . . . . . . . 241
    Assigned Job 27-4—Welding a Lap Joint in the Flat Welding Position . . . . . . . 243
    Assigned Job 27-5—Welding a T-Joint in the Flat Welding Position . . . . . . . 245
    Assigned Job 27-6—Welding a Square-Groove Butt Joint in the Flat Welding Position . . . . . . . . . . . . . . . . . . . 247

**Lesson 28**—Oxyfuel Gas Welding: Horizontal, Vertical, and Overhead Welding Positions . . . 249
    Assigned Job 28-1—Welding a Lap Joint in the Horizontal Welding Position . . 251
    Assigned Job 28-2—Welding a T-Joint in the Horizontal Welding Position . . 253
    Assigned Job 28-3—Welding a Square-Groove Butt Joint in the Horizontal Welding Position . . . . . . . . . . . . . 255
    Assigned Job 28-4—Welding a Lap Joint in the Vertical Welding Position . . . . 257
    Assigned Job 28-5—Welding a T-Joint in the Vertical Welding Position . . . . 259
    Assigned Job 28-6—Welding a Square-Groove Butt Joint in the Vertical Welding Position . . . . . . . . . . . . . 261
    Assigned Job 28-7—Welding a Lap Joint in the Overhead Welding Position . . 263
    Assigned Job 28-8—Welding a T-Joint in the Overhead Welding Position . . 265
    Assigned Job 28-9—Welding a Square-Groove and V-Groove Butt Joint in the Overhead Welding Position . . . . . 267

**Lesson 29**—Brazing and Braze Welding . . . 269
    Assigned Job 29-1—Brazing a Lap Joint on Thin Mild Steel in the Flat Position . . 273
    Assigned Job 29-2—Braze Welding an Outside Corner Joint in the Flat Welding Position . . . . . . . . . . . . . 275

**Lesson 30**—Soldering . . . . . . . . . . . . . . . . 277
    Assigned Job 30-1—Soldering Copper Pipe Fittings . . . . . . . . . . . . . . . . . 281

## Section 7
# Resistance Welding

Resistance Welding Safety Test . . . . . . . . . . 283

**Lesson 31**—Resistance Welding: Equipment and Supplies . . . . . . . . . . . . . . . . . . . . . . . 285
    Assigned Job 31-1—Identifying the Parts of a Resistance Welding Machine . . . . 287

**Lesson 32**—Resistance Welding: Procedures . 289
    Assigned Job 32-1—Adjusting the Resistance Welding Machine . . . . 291
    Assigned Job 32-2—Making Resistance Spot Welds in Mild Steel . . . . . . . 293
    Assigned Job 32-3—Practicing Resistance Spot Welding Mild Steel . . . . . . . . 295
    Assigned Job 32-4—Making Resistance Spot Welds in Aluminum . . . . . . . 297

## Section 8
# Welding in Industry

**Lesson 33**—Welding Pipe and Tube . . . . . . 299
    Assigned Job 33-1—Welding a V-Groove Butt Joint on Pipe in the Horizontal Rotated (1G) Position . . . . . . . . . . 303
    Assigned Job 33-2—Welding a V-Groove Butt Joint on Pipe in the Horizontal Fixed (5G) Position . . . . . . . . . . . 307

**Lesson 34**—Robotics in Welding . . . . . . . . 311

**Lesson 35**—Special Welding and Cutting Processes . . . . . . . . . . . . . . . . . . . . . . . . 315

**Lesson 36**—Inspecting and Testing Welds . . 317

**Lesson 37**—Welder Certification . . . . . . . . 319

# Section 1
## Introduction to Welding

### Lesson 1
### Careers in Welding

### Objectives:
You will be able to choose, prepare for, gain, and keep an appropriate job for yourself in the welding industry whether you become an entrepreneur or employee.

### Instructions:
Carefully read Chapter 1 of the text and study Figures 1-1 through 1-12. Then, answer or complete the following questions.

1. As you set your career goals, what three factors about yourself should you consider?

   _____

   _____

   _____

2. The minimum level of education for a career as a welding inspector is ____ ____ or ____ ____.

   _____

3. What is the minimum level of education for a person to be hired as a metallurgist?

   _____

_____    4. Which of the following high school courses is *not* critical for success in the welding industry?
   A. Geometry.
   B. Computer.
   C. Language arts.
   D. History.

5. Why are communication skills important in the welding workplace?

   _____

   _____

6. According to the text, what does ethical behavior mean?

   _____

   _____

_____    7.  Based on Figure 1-6, which of the following are ways you can demonstrate self-confidence?
A. Speak with enthusiasm and confidence. Speak clearly and concisely.
B. Seek leadership positions in school, clubs, church, and other organizations. Be honest.
C. Be sure you understand your task or assignment, then perform it in a calm and relaxed manner. Show pride in your work.
D. Follow written and verbal directions. Demonstrate loyalty to your school and superiors.

8.  What are three documents that can help you get a good job in the welding industry? (Hint: They all start with the letter R!)

R_____

R_____

Letter of R_____

9.  List any four of the pitfalls to avoid when applying for a welding job.

_____

_____

_____

_____

_____    10.  Which of the following welders would be most likely to succeed as an entrepreneur?
A. Dale is a hard worker and can put in long hours. He just graduated from technical college and has some debt and no one in his family has money to invest in his new business.
B. Scott has been working as a welder and has received many commendations from his boss. He has identified a need in the industry for a specialized process that he enjoys and is skillful at doing. His parents have offered to invest in his new business.
C. Paul lives with his girlfriend and their two children. He also takes care of his child from a previous marriage every other weekend and pays child support. He wants to open his own business so he can spend more time with his family.
D. Tom does not like waking up early. He does not get along well with his boss so he wants to be his own boss. On weekends, he likes to go fishing with his friends. He has not made many friends at work though, because they do not like that he is always late getting back to work after lunch and lets them do the work.

Name: _____  Date: _____

Class: _____  Instructor: _____

Lesson Grade: _____  Instructor's Initials: _____

## Lesson 2
# Safety in the Welding Shop

## Objectives:

You will be able to identify the hazards that exist in the welding shop. You will be able to list the general safety rules regarding the proper storage of compressed gas . You will be able to protect yourself and your coworkers by knowing the latest safety procedures and health information related to the field of welding.

## Instructions:

Carefully read Chapter 2 of the text and study Figures 2-1 through 2-14. Then, answer or complete the following questions.

1. *True or False?* Welding helmets with filter lenses are designed to protect welders from most hazards that welders face, including arc rays, weld sparks, slag chips, and grinding fragments.

2. What shade number should be used in the welding helmet when performing shielded metal arc welding with an arc current of 200 amps?

   _____

3. Name at least three safety precautions that must be taken to avoid serious injury from a compressed gas cylinder leaking or exploding.

   _____

   _____

   _____

_____  4. Which type of respirator delivers a constant flow of clean breathing air to the welder rather than filtering contaminated air?
   A. Negative-pressure air-purifying respirator.
   B. Positive-pressure respirator.
   C. $CO_2$ scrubbing rebreather.
   D. All of the above.

5. Many welding areas have excessive noise, which is dangerous to a worker's hearing. What is the name of the organization that sets limits to the amount of noise exposure that is acceptable?

   _____

6. What protection should be worn if a welder is working in an area that exceeds the set limits for noise?

   _____

7. What danger to the lungs do welders face when working in confined spaces?

   _____

   _____

8. What precautions should be taken to ensure the air that welders breathe is clean and safe?

   _____

   _____

9. List five pieces of protective clothing and safety gear that welders should wear.

   _____

   _____

   _____

   _____

   _____

10. What is the name of the document that covers the safety precautions and procedures for working with containers that have held hazardous or flammable materials?

    _____

11. To quench a small wood fire on a job site, which class fire extinguisher would you use?

    _____

# Lesson 3
# Welding and Cutting Processes

## Objectives:
You will be able to describe some of the uses of welding and the development of several welding processes. You will give reasons for choosing some welding processes over other welding or joining processes.

## Instructions:
Carefully read Chapter 3 of the text and study Figures 3-1 through 3-9. Then, answer or complete the following questions.

1. *True or False?* Blacksmiths are no longer employed by modern industries.

_____ 2. Which of the following describes the first models of parts that later may be produced for sale?
   A. Prototype.
   B. Machined.
   C. Weldments.
   D. Small lots.

3. How many different welding and cutting processes are defined by the AWS? _____

4. *True or False?* Arc welding processes can be used to weld all base metals.

5. What fuel gas is most commonly combined with oxygen in oxyfuel welding?

   _____

6. Industrial operations such as riveting, hammering, welding, forging, casting, turning, and sawing are known as _____.

   _____

7. What creates the heat in resistance welding?

   _____

   _____

8. In what year were oxyfuel gas welding torches developed?

   _____

9. In what year was welding first done on copper?

_____

10. Define welding.

_____

_____

_____

_____

11. List four recently developed welding processes.

_____

_____

_____

_____

12. Stones can be welded together! Explain how this can be done.

_____

_____

_____

13. In what year was the first all-welded building constructed?

_____

_____  14. Metals can be welded by which of the following methods?
   A.  Forge welding.
   B.  Explosion welding.
   C.  Cold welding.
   D.  All of the above.

15. Many specialized welding and cutting processes are affordable only to large _____ companies.

_____

# Lesson 4
# The Physics of Welding

## Objectives:

You will be able to explain the effects of heat, pressure, expansion, and contraction on welding. You will understand and be able to describe the chemical and mechanical properties of metal. You will be able to describe the effects of heat treating on a metal's properties. You will understand the electrical relationship between voltage and current. You will be able to use and convert US customary and SI metric measurement units.

## Instructions:

Read Chapter 4 in the text and study Figures 4-1 through 4-9. Then, answer or complete the following questions.

1. In any arc welding, heat is created by an arc that is struck between a(n) _____ and the _____ _____.

_____

_____     2. In which direction, A or B, will the vertical piece tend to move when the following joint is welded?

A ←———————     ———————→ B

_____     3. Which of the following can be used to relieve stress and/or prevent movement during welding?
      A.   Tack welding.
      B.   Align the parts to allow for contraction during welding.
      C.   Preheating the parts.
      D.   All of the above.

4. List two physical properties of a metal.

_____

_____

5. What term describes a material's ability to resist pulling forces?

_____

6. Explain how the resistance spot welding process uses both heat and pressure to form a weld.

_____

_____

7. The properties of a metal are largely determined by its _____ composition (the elements that make up the metal).

_____

_____    8. As a metal is cooled, its natural tendency is to ____.
       A. expand
       B. crack
       C. contract
       D. lose toughness

9. A material with very low ductility is _____.

_____

10. A material is said to be ____ if it can resist scratching and indenting.

_____

11. The process of quickly cooling a heated metal by immersing it in a liquid is called ____.

_____

12. The process of heating a metal and then allowing it to slowly cool in the oven is called ____.

_____

_____    13. Interpass heating ____ an elevated temperature during welding.
       A. decreases
       B. maintains
       C. increases

14. *True or False?* Open-circuit voltage is measured while the circuit is closed and current is flowing.

_____    15. Current flow, or the number of electrons flowing in a closed electric circuit, is measured in ____.
       A. ohms
       B. volts
       C. amperes
       D. pounds

_____    16. The process of heating a metal and then removing it from the oven to slowly return to room temperature is called ____.
       A. stress relieving
       B. interpass heating
       C. annealing
       D. normalizing

_____    17. The force that causes electrons to flow in an electric circuit is called ____.
       A. voltage
       B. amperage
       C. resistance
       D. impedance

18. What is the purpose of preheating and interpass heating?

_____

_____

19. In the US customary system, flow rate is measured in ____.

_____

20. *True or False?* The arc gap between the electrode and the base metal offers no resistance to the flow of electrons.

Name: _____  Date: _____

Class: _____  Instructor: _____

Lesson Grade: _____  Instructor's Initials: _____

## Lesson 5
# Math for Welding

## Objectives:

After studying this chapter, you will understand how the STEM subjects apply to welding. You will also confidently add, subtract, multiply, and divide whole numbers, fractions, and decimals with and without a calculator.

## Instructions:

Carefully read Chapter 5 of the text and study Figures 5-1 through 5-7. Then, perform the following calculations. Do not use a calculator unless directed to do so by your instructor.

1. $78 + 24$

   _____

2. $125 + 36 + 16$

   _____

3. $180 - 72$

   _____

4. $6582 - 27$

   _____

5. $84 + 18 - 60$

   _____

6. $68 \times 8$

   _____

7. $23 \times 5$

   _____

8. $36 \div 12$

   _____

9. $5/8 + 1/4$

   _____

10.  2 1/4 + 5 1/2

_____

11.  3.4 + 5.78

_____

12.  8.5 × 2.4

_____

13.  8.21 + 9.024

_____

14.  9.845 − 2.05

_____

15.  3/5 × 4/8

_____

16.  1/8 ÷ 1/2

_____

17.  2 1/2 × 4 1/4

_____

*Convert the following from fractions to decimals or from decimals to fractions.*

18.  1/4

_____

19.  5/8

_____

20.  .8

_____

*Calculate the following and show your work.*

21.  A welder earns $18.50 per hour. How much does the welder earn for 40 hours of work?

_____

_____

22.  A repair piping job in a factory is going to take 100 hours of work. The work will be done in eight hours on a Saturday. How many workers are needed to complete the work in the time allowed? Round up to the next whole number.

_____

_____

Name _____

23. Four friends go to dinner and order a large pizza and four drinks. The bill comes to $29.84. How much does each person have to pay? Show the actual amount, then round up to the next tenth of a dollar.

   _____

   _____

24. A square steel tube weighs 4 pounds per foot. How much will 10 pieces that are each 8 feet long weigh?

   _____

   _____

25. A welder using the gas metal arc welding process sets the shielding gas flow rate to 30 cubic feet per hour. How many cubic feet of gas will the welder use in 4 hours of continuous welding?

   _____

   _____

Name: _____  Date: _____

Class: _____  Instructor: _____

Lesson Grade: _____  Instructor's Initials: _____

# Lesson 6
# Math Applications for Welders

## Objectives:

After studying this chapter, you will be able to use the US customary system and the SI metric system to measure and convert values. You will also be able to calculate the perimeter, area, and volume of many common shapes.

## Instructions:

Carefully read Chapter 6 of the text and study Figures 6-1 through 6-20. Then, answer or complete the following sentences.

1. Most of the world uses which system of measurement?

   _____

2. Explain the difference between acute, obtuse, right, and equilateral triangles.

   _____

   _____

   _____

   _____

3. What is the sum of the angles in a triangle?

   _____

4. How many millimeters are equal to 8"? Round to a whole number.

   _____

5. How many inches are equal to 815 mm? Round to a whole number.

   _____

*For questions 6–10, use the following information. A welder must repair a section of schedule 40 piping. The pipe diameter has an 8.0" inside diameter and the pipe wall is .322" thick. A 3' section of pipe is removed and a replacement section is welded in place. Both ends of the replacement pipe must be welded.*

6. What is the radius of the pipe?

   _____

7. How many inches of weld are required to weld one pass around the pipe? Round to nearest tenth inch. Hint: What is the circumference of an 8″ circle?

_____

8. Typically four weld passes are required to weld this type of weld joint in the horizontal position. How many inches of weld are required to complete the entire weld on one end of the pipe? Hint: A welder will need to weld the circumference four times. Round up to the nearest tenth inch.

_____

9. A pipe in the horizontal position will be welded from the bottom to the top. One shielded metal electrode will make a weld about 3″ long. How many electrodes are required to complete all four passes on the weld joints on both ends of the pipe? Round up to the next whole number.

_____

10. What is the volume of this 3′ section of pipe in cubic inches?

_____

_For questions 11–15, a welder is making a ramp for a loading dock in a building wall. The dock is 4′ above ground level (4′ tall). The wall meets the ground at a 90° angle. The dock and ramp are 8′ wide._

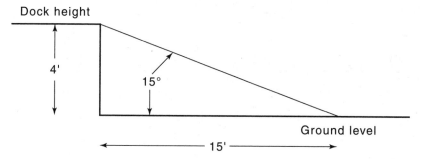

11. The ramp will meet the ground at a 15° incline. At what angle does the ramp meet the dock?

_____

12. If the bottom of the ramp starts 15′ out from the ground at the dock, how long does the ramp need to be to reach the top of the dock? (Use the Pythagorean Theorem and round your answer to the nearest tenth.)

_____

_____

13. What is the surface area of the ramp?

_____

14. If triangle metal sides are welded to the sides of the ramp for support, what is the area of each side piece?

_____

Name _____

15. If rectangular pieces of 4′ × 8′ steel plate are used to form the surface of the ramp, how many pieces would be needed to complete the ramp surface? Round up to the next whole number. (Note that one of the pieces would need to be cut on one side to fit properly.)

_____

_____

16. If a potential customer wants to know how much a job will cost, you would submit a(n) _____.

_____

17. In determining the estimated cost of a welding task, what three costs will always be included?

_____

_____

_____

18. What are four other costs that may be considered in a welding project?

_____

_____

19. Explain the difference between base materials and welding materials.

_____

_____

_____

20. A temperature of 65°F is equal to how many degrees Celsius (°C)? Use the formula °C = (°F − 32) ÷ 1.8. Show your work in the space provided.

_____

_____

_____

_____

Name: _____     Date: _____

Class: _____     Instructor: _____

Lesson Grade: _____     Instructor's Initials: _____

# Lesson 7
# Weld Joints and Positions

## Objectives:

You will be able to identify the five basic weld joints, the types of welds used on each joint, and the parts or areas of a welded joint. You will be able to define the terms used to describe welds, joint geometry, and the positions used in welding.

## Instructions:

Read Chapter 7 and carefully study Figures 7-1 through 7-20. Then, answer or complete the following questions.

1. Name the parts of the fillet weld shown in the following figure:

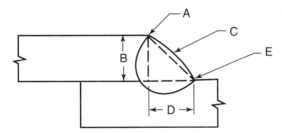

A. _____

B. _____

C. _____

D. _____

E. _____

2. Name the joints shown in the following figure:

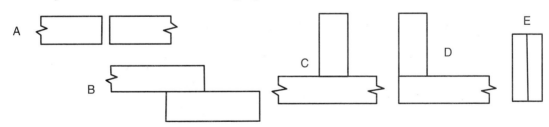

A. _____     D. _____

B. _____     E. _____

C. _____

3. What type of corner joint is shown in the following figure? _____

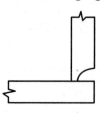

4. In the T-joint shown in the following figure, two welds were made. The first was a(n) ____-groove weld. The second weld was a(n) ____ weld.

_____

_____

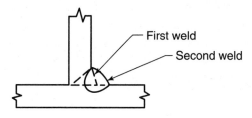

5. List three types of weld joints that can be welded without the addition of filler material.

_____

_____

_____

6. What type of butt joint is shown in the following figure?

_____

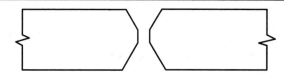

7. Name the various parts of the single-V-groove welded butt joint shown in the following figure:

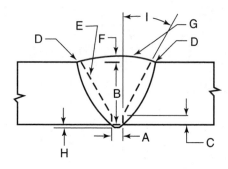

A. _____

B. _____

C. _____

D. _____

E. _____

F. _____

G. _____

H. _____

I. _____

Name _____

8. What type of butt joint is shown in the following figure?

   _____

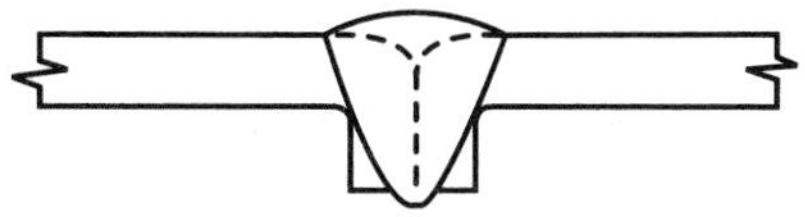

9. One or more weld beads, or weld ____, may be needed to complete a weld on thick metal.

   _____

10. What type of butt joint is shown in the following figure?

    _____

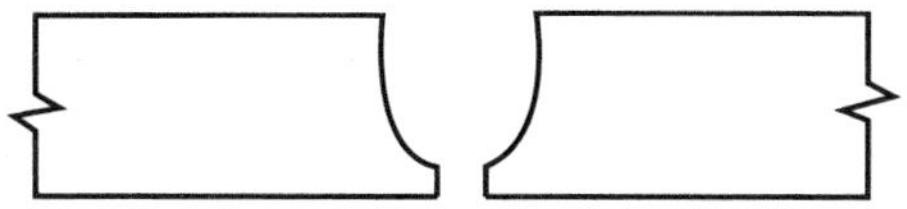

11. *True or False?* If the weld groove is too large, the welder's time and additional filler metal will be wasted.

12. What type of weld bead is being laid in the following drawing?

    _____

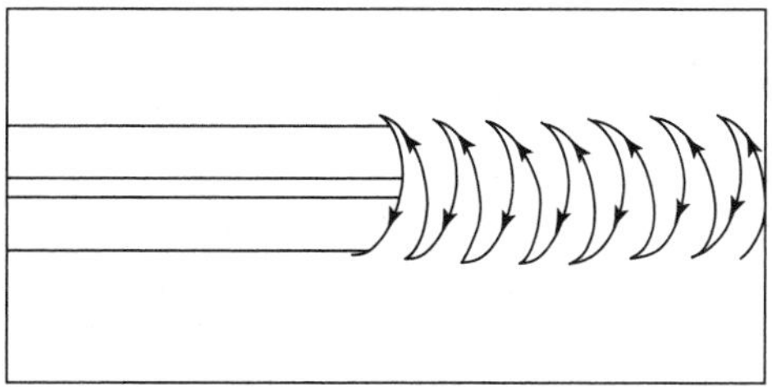

_____   13. If the weld axis is within 15° of horizontal and the weld face is between 80°–150° or 210°–280° (angles are measured with 0° as the straight down position), the weld is being done in ____.
   A. the horizontal welding position
   B. the vertical welding position
   C. the flat welding position
   D. the overhead welding position
   E. None of the above.

14. Define the term *joint geometry*.

    _____

    _____

_____   15. Which of the items listed below is *not* normally used to hold a weld joint to prevent it from warping or changing shape during welding?
   A. Clamps.
   B. Jigs.
   C. Fixtures.
   D. Slip-joint pliers.

16. Fill in the blanks to describe the 4G welding position in detail.
    A.  The weld axis is between ____° and ____°.
    B.  The weld face is between ____° and ____° or ____° and ____°.
    C.  The weld is made from the _____ side of the joint.

17. Define the term *weld axis*.

    _____

    _____

_____    18.  Which of the following methods is *not* normally used to prepare the edges of a weld
               joint prior to welding?
               A.  Machining.
               B.  Chiseling.
               C.  Gouging.
               D.  Flame cutting.

19. In the following drawing, in what position is the joint being welded? _____

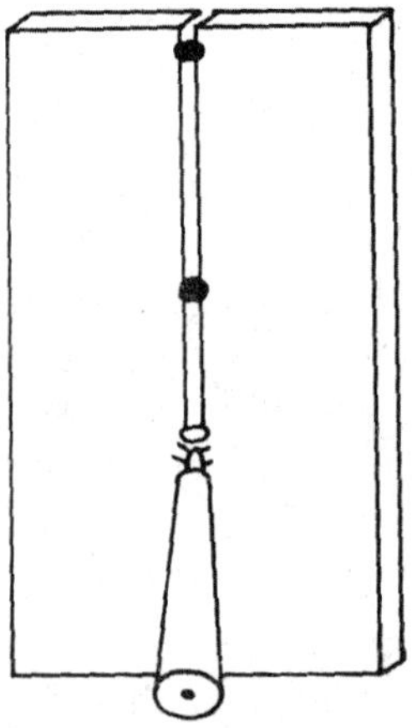

20. Fill in the blanks with the number and letter combination that the American Welding Society
    has assigned to groove welds in each of the welding positions.

    A.  Flat welding position:_____

    B.  Overhead welding position: _____

    C.  Vertical welding position:_____

    D.  Horizontal welding position: _____

Name: _____   Date: _____

Class: _____   Instructor: _____

Lesson Grade: _____   Instructor's Initials: _____

# Lesson 8
# Welding Symbols

## Objectives:

You will be able to identify lines and points in a three-view drawing of a part. You will also be able to describe the various information given on an American Welding Society welding symbol.

## Instructions:

Read Chapter 8 and study Figures 8-1 through 8-27. Then, answer or complete the following questions.

_____   1.  Drawings used to produce parts in a factory are called mechanical drawings because _____.
   A.  they illustrate mechanical parts
   B.  they show architectural parts
   C.  they show welded parts
   D.  they were traditionally made with the use of instruments by mechanical means
   E.  All of the above.

2.  The _____ typically contains information such as the permitted part size tolerances, drawing scale, and material the part is to be made of.

_____

3.  *True or False?* In the orthographic drawing process, eight regular views of a part can be shown.

4.  Name the views shown in the three-view drawing for Questions 5–10.

_____

_____

_____

*For Questions 5–10, match the lines in the three-dimensional view with the same lines in the three-view orthographic drawing. Remember that a line in one view may appear as a point in another view. Not all letters will be used.*

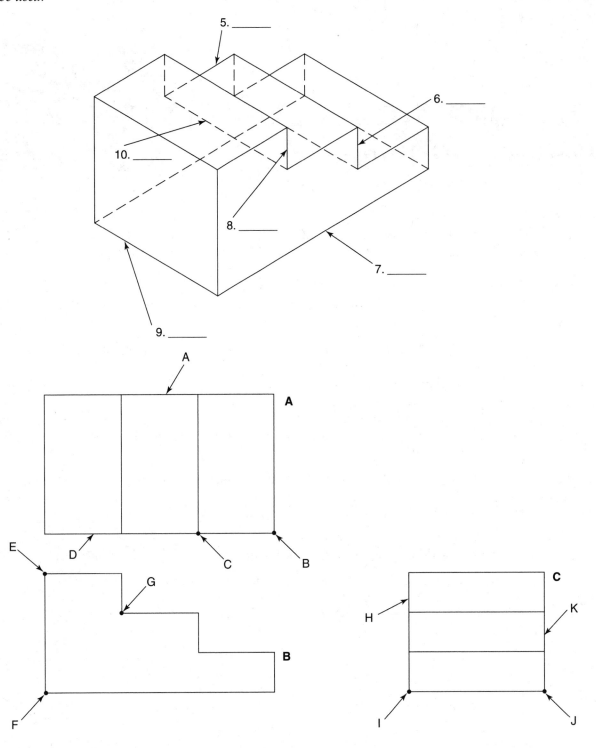

Lesson 08   *Welding Symbols*   **29**

Name _____

*For Questions 11–15, name the type of weld to be made on each joint, as shown by the AWS weld symbols.*

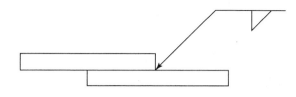

11. _____ weld on a lap joint.

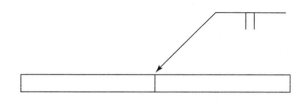

12. _____-_____ _____
    weld on a butt joint.

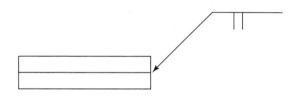

13. _____-_____ _____
    weld.

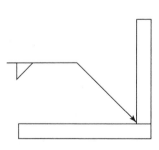

14. _____ weld on a T-joint.

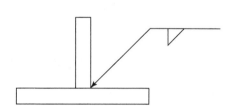

15. _____ weld on an inside corner
    joint.

16. *True or False?* On a mechanical drawing of a weldment, the basic weld joint is shown by the position of the metal parts.

17. *True or False?* Information must always be shown in the same area on the AWS welding symbol.

18. Why is a tail sometimes used on a welding symbol?

_____

_____

_____

Copyright Goodheart-Willcox Co., Inc. May not be reproduced or posted to a publicly accessible website.

*For Questions 19–22, match each AWS welding symbol on the left with the appropriate weld cross-section on the right.*

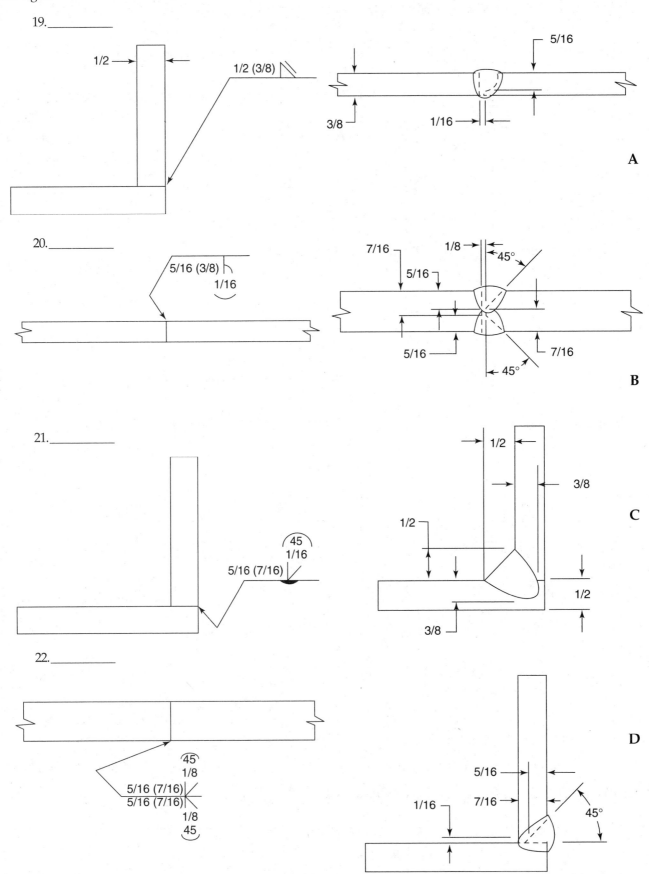

19._____

20._____

21._____

22._____

Name _____

_____ 23.   Which of the following statements is *not* true?
   A.   The reference line is always drawn horizontally.
   B.   The welding symbol must appear in every view on the drawing where the weld is shown.
   C.   The arrow may be drawn from either end of the reference line.
   D.   The arrow may be bent to show which part of a weldment is to be prepared for welding.
   E.   The arrowhead must touch the line to be welded.

24.   Identify the arrow side and the other side in the following drawing.

   A._____

   B._____

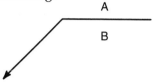

_____ 25.   Which of the following is *not* true for the root opening shown on a welding symbol?
   A.   The root opening size is given just outside the basic weld symbol.
   B.   The root opening may be given as a fraction, decimal, or metric dimension, on a welding symbol.
   C.   The root opening is the space between the base metals at the root of the joint.
   D.   The root opening may be held constant by a jig or fixture without tack welding.

26.   *True or False?* The edge angle cut on each piece of base metal in a V-groove butt joint is one-half the angle shown on the welding symbol.

27.   What welding process is used to make the weld indicated in the following drawing?

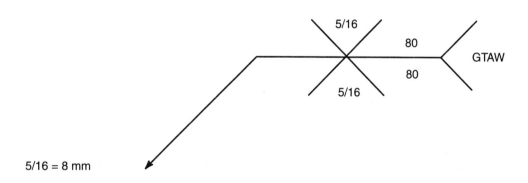

_____ 28.   What size or dimension is always placed in parentheses?
   A.   The root opening.
   B.   The weld size.
   C.   The effective throat.
   D.   The pitch.
   E.   The length.

29. According to the AWS welding symbol in the following drawing, the contour of the weld is ____. The weld is to be finished by ____.

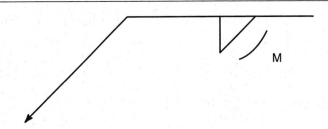

30. Answer the following questions about the drawing shown here.

A. How deeply would you grind the metal for the weld shown in the drawing?

_____

B. At what angle would you grind each piece of base metal to prepare the joint for this weld?

_____

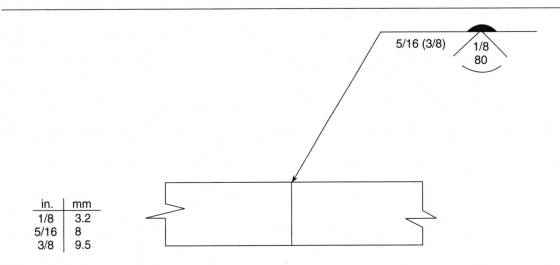

5/16 (3/8)    1/8    80

| in. | mm |
|-----|-----|
| 1/8 | 3.2 |
| 5/16 | 8 |
| 3/8 | 9.5 |

_____ 31. According to the AWS welding symbol in the following drawing, the weld is ____.
  A. continuous
  B. intermittent
  C. elevated
  D. inverted

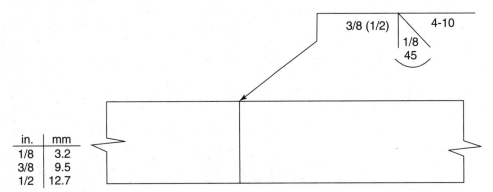

3/8 (1/2)    4-10    1/8    45

| in. | mm |
|-----|-----|
| 1/8 | 3.2 |
| 3/8 | 9.5 |
| 1/2 | 12.7 |

Name _____

32. *True or False?* In the following drawing, the completed weld on the right is correctly made according to the AWS welding symbol on the left.

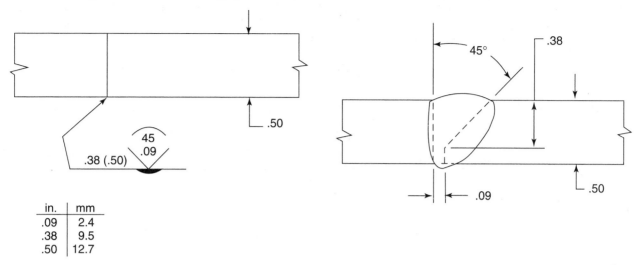

| in. | mm |
|-----|------|
| .09 | 2.4 |
| .38 | 9.5 |
| .50 | 12.7 |

33. *True or False?* The melt-through symbol is primarily used for joints that are welded from both sides.

_____ 34. Which of the following statements regarding the AWS welding symbol is *not* true?
   A.   The reference line nearest the arrowhead indicates the first operation.
   B.   The backing weld symbol is used when a stringer bead is laid on the root side of the weld to ensure complete penetration.
   C.   Weld root opening size is given inside the basic weld symbol.
   D.   The field weld symbol is used to indicate the polarity that should be used to make the weld.

35. A(n) ____ ____ is made through a hole that is not round.

Name: _____   Date: _____

Class: _____   Instructor: _____

Lesson Grade: _____   Instructor's Initials: _____

## Assigned Job 8-1
# Reading an AWS Welding Symbol

## Objectives:

You will learn to recognize and interpret common AWS welding symbols. You will also learn how to obtain all required information from an AWS welding symbol.

1. Answer the following questions based on the information provided in the welding symbol:

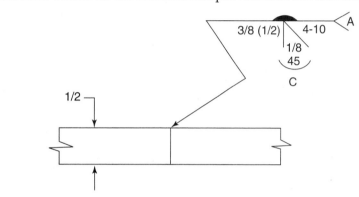

| in. | mm |
|-----|------|
| 1/8 | 3.2 |
| 3/8 | 9.5 |
| 1/2 | 12.7 |

Note A: GMAW, .030 wire.
Use company welding code #24

A. Type of joint: _____

B. Weld root opening size: _____

C. Total groove angle: _____

D. Beveled piece (left or right): _____

E. Weld size: _____

F. Welding process: _____

G. Welding wire diameter: _____

H. Welding code: _____

I. Weld groove depth: _____

J. Continuous or intermittent weld: _____

K. Weld length: _____

L. Pitch of welds: _____

M. Bead contour shape: _____

N. Weld finish: _____

O. What does the melt-through symbol indicate? _____

2. Sketch and dimension the weld described by this welding symbol:

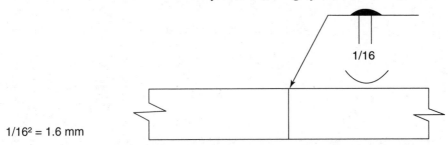

1/16

$1/16^2 = 1.6$ mm

3. Answer the following questions from the information given on the welding symbol:

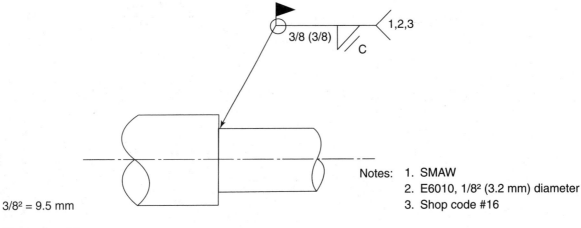

3/8 (3/8)    C    1,2,3

$3/8^2 = 9.5$ mm

Notes:  1. SMAW
        2. E6010, $1/8^2$ (3.2 mm) diameter
        3. Shop code #16

A.  Type of weld: _____

B.  Weld size: _____

C.  Effective throat size: _____

D.  Continuous or intermittent weld: _____

E.  Bead contour shape: _____

F.  Weld finish: _____

G.  Welding process: _____

H.  Size and type electrode: _____

I.  Welding code: _____

J.  The circle means: _____

K.  Location where weld is performed: _____

Name _____

4.  Sketch and dimension the size, shape, and effective throat of the weld described by this welding symbol:

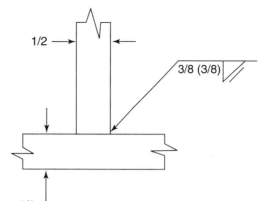

| in. | mm |
|-----|------|
| 3/8 | 9.5 |
| 1/2 | 12.7 |

## Inspection:

Recheck the welding symbols against the weld information to ensure that all the answers and drawings are complete and correct.

Name: _____    Date: _____

Class: _____    Instructor: _____

Lesson Grade: _____    Instructor's Initials: _____

## Assigned Job 8-2
# Drawing an AWS Welding Symbol

## Objective:

You will learn to draw the appropriate welding symbol to completely describe a desired weldment.

1.  In the figure provided on the right, draw the correct welding symbol to describe the weld shown on the left. Indicate in the welding symbol that complete penetration is required.

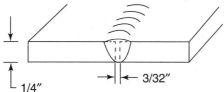
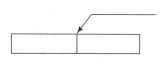

2.  In the figure provided on the right, draw the correct welding symbol to describe the weld shown on the left. Indicate in the welding symbol that the GMAW process should be used.

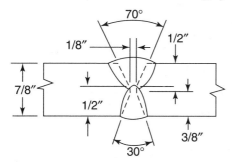
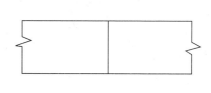

3.  In the figure provided on the right, draw the correct welding symbol to describe the weld shown on the left. Indicate in the welding symbol that the SMAW process should be used.

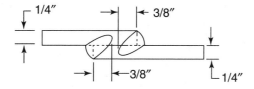
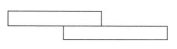

4.  In the figure provided on the right, draw the correct welding symbol to describe the weld shown on the left. Indicate in the welding symbol that the welds should be 1/4″ fillet welds 3/8″ deep.

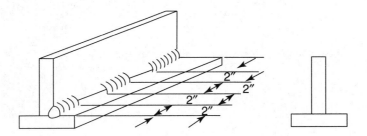

5.  In the figure provided on the right, draw the correct welding symbol to describe the weld shown on the left. Indicate in the welding symbol that the same weld is to be made on all edges of the weldment.

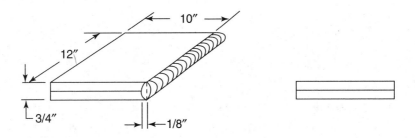

## Inspection:

Recheck the welding symbols against the weld information to ensure that all of your answers are complete and correct.

Name: _____   Date: _____

Class: _____   Instructor: _____

Lesson Grade: _____   Instructor's Initials: _____

# Section 2
# Shielded Metal Arc Welding

## Shielded Metal Arc Welding Safety Test

### Objectives:

You will be able to correctly assemble a SMAW outfit. You will also be able to describe the safety precautions that must be observed when working with SMAW equipment and welding with the shielded metal arc.

### Instructions:

Always follow safe practices when welding. If you have safety questions or concerns, ask your instructor. This test does not include questions about every safety topic, but is intended to highlight key items. Review the material in Chapters 9, 10, and 11 and read the SMAW Safety Precautions section in Chapter 12. Then, answer or complete the following questions.

_____   1.   The arc in SMAW can reach temperatures over ____.
A.   2800°F (1540°C)
B.   3590°F (1980°C)
C.   5000°F (2760°C)
D.   9000°F (4980°C)
E.   11000°F (6090°C)

2.   *True or False?* A welding helmet will keep a welder's eyes safe from all shop hazards. Additional safety goggles are not required.

3.   *True or False?* The polarity switch should never be changed while you are welding.

4.   What part of a complete SMAW welding station is missing from the following list?

---

A chipping hammer and wire brush.

An arc welding machine.

An electrode holder.

Covered electrodes.

A welding table.

Electrode and work leads.

A helmet and lenses.

A welding booth or screen.

_____ 5. Which type of arc welding machine is best for manual arc welding?
   A.   AC.
   B.   DC.
   C.   CC.
   D.   CV.
   E.   CB.

_____ 6. Which term below tells the welder how long the welding machine can be used during each 10-minute period?
   A.   Open circuit voltage.
   B.   Duty cycle.
   C.   Rated output current rating.
   D.   Input power requirements.
   E.   Welding allowance.

7. The maximum open circuit voltage for most manual AC or DC machines is _____V. This relatively low voltage protects workers from electrical shock.

_____

8. *True or False?* The handle of the electrode holder is well-insulated to protect the welder from electrical shock.

9. Placing a hand and wet glove on the worktable or electrode while welding may cause the welder to receive a(n) _____ _____.

_____

10. *True or False?* If you will be performing shielded metal arc welding for 30 seconds or less, flash goggles with #3 filter lenses provide adequate eye protection.

_____ 11. It is advisable to have a(n) _____ if you are shielded metal arc welding in an area where a fire may occur.
   A.   fire watch
   B.   smoke alarm
   C.   safety interlock
   D.   bucket of water close at hand
   E.   All of the above.

12. *True or False?* Depending on the size of the electrode or amperage used, a #10–#14 filter lens is used during SMAW.

_____ 13. What should welders wear to protect their eyes from arc flashes from behind the arc shield?
   A.   A large cap.
   B.   A leather cape.
   C.   A high collar.
   D.   Flash goggles with a #1–#3 lens.
   E.   Goggles or glasses with a #10 lens.

_____ 14. When welding is being performed in the overhead position, it is best to place the exhaust pickup tube so that fumes are drawn to a point _____ the welder's head in order to protect the welder from toxic fumes.
   A.   above
   B.   below
   C.   behind

15. *True or False?* Always wear chipping goggles when chipping the slag from a weld.

Name _____

16. Welding leads must be well ____ to prevent electrical shock.

_____

17. *True or False?* The electrode or electrode holder must be in contact with the workpiece or work-table when the arc welding machine is turned on.

18. To prevent electrical shock, the power to the arc welding machine must be ____ ____ whenever a safety check of the welding station is made.

_____

_____ 19. The walls or curtains of the arc welding booth protect ____ from arc flashes and flying slag.
   A.  the welder
   B.  the welding machine
   C.  nearby workers
   D.  the weld
   E.  All of the above.

20. The polarity of a DC welding machine can be changed by moving the polarity switch or by changing the ____ lead and the ____ lead.

_____

21. *True or False?* Coarse amperage range adjustments may be made by moving a lever on the face of the machine or by changing the lead positions in the jacks on the machine.

22. When arc welding in any position, the collar on the welder's coveralls or shirt must be ____.

_____

23. *True or False?* Gauntlet-type welding gloves are the best type to use when welding out-of-position.

_____ 24. ____ make current changes with the electrode touching the workpiece, worktable, or while welding.
   A.  Always
   B.  Never

_____ 25. Which of the materials listed below may produce toxic (harmful) fumes when welding is done on or near it?
   A.  Cadmium.
   B.  Zinc.
   C.  Lead.
   D.  Chlorinated hydrocarbons.
   E.  All of the above.

Name: _____    Date: _____

Class: _____    Instructor: _____

Lesson Grade: _____    Instructor's Initials: _____

# Lesson 9
# SMAW: Equipment and Supplies

## Objectives:

You will be able to describe the types of current used in the SMAW process, as well as the types of equipment and supplies that are required for SMAW.

## Instructions:

Read Chapter 9 and study Figures 9-1 through 9-30. Then, answer or complete the following questions.

1. *True or False?* In the electron flow theory, electrons flow from the positively charged body to the negatively charged body.

2. Name or describe four methods of attaching the welding leads to the welding machine or base metal.

   _____

   _____

_____    3.  When the electrons flow from the base metal to the electrode, the flow is called ____.
   A.  direct current electrode negative
   B.  alternating current electrode negative
   C.  alternating current electrode positive
   D.  direct current electrode positive
   E.  direct current straight polarity

4. Look at the following figure. What is the recommended duty cycle when welding with 120 amperes?

   _____

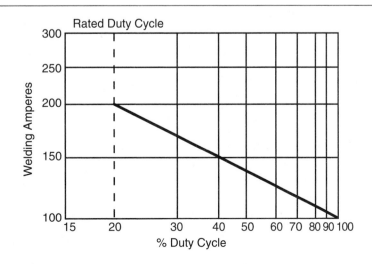

5. What number filter lens is recommended when welding with SMAW and using 1/8″ (3.2 mm) covered electrodes?

_____

6. Name the items that make up an SMAW outfit and station.

   **Outfit:**

   _____

   _____

   **Station (all of the above plus):**

   _____

   _____

7. *True or False?* In shielded metal arc welding, the arc is shielded from the atmosphere by a gas that is created as a covered electrode melts.

_____     8. With _____, the electron flow changes direction many times each second.
   A.  direct current
   B.  alternating current

9. *True or False?* DCEN is the same as the old electron flow designation of direct current straight polarity.

_____     10. What type of arc welding machine is best for manual arc welding?
   A.  Constant current (CC).
   B.  Constant voltage (CV).

11. A welding machine that is called a "drooper" is a(n) _____ machine.

_____

12. Name four variables that must be considered when selecting an arc welding machine.

_____

_____

13. Describe two ways of changing the polarity of the circuit on a direct current machine.

_____

_____

_____     14. The maximum OCV for most arc welding machines is _____.
   A.  40V
   B.  60V
   C.  80V
   D.  100V
   E.  115V

15. What size welding lead is recommended for carrying 150 amperes to a job 100 feet away from the arc welding machine? Refer to Figure 9-18 in the text.

_____

Name _____

16. A CV arc welding machine produces a(n) _____ that does not change very much from the _____ that was set.

_____

_____ 17. Which is the healthiest location to position a flexible exhaust pickup tube?
  A.  On the workbench surface while doing an overhead weld.
  B.  Above the welder's head when welding in the flat welding position.
  C.  Below the welder's face when welding in the flat welding position.
  D.  As far from the welder as possible.

_____ 18. In the U.S.A., how long does alternating current generally flow in one direction per cycle?
  A.  1/60 second.
  B.  1/100 second.
  C.  1/120 second.
  D.  1/80 second.
  E.  1/30 second.

_____ 19. The temperature created by the arc in SMAW is _____.
  A.  6300°F–6500°F
  B.  6500°C–7000°C
  C.  4500°C–5000°C
  D.  2500°F–2700°F
  E.  2500°C–2700°C

20. *True or False?* In DCEP, the electrons travel from the electrode to the base metal.

# Lesson 10
# SMAW: Equipment Assembly and Adjustment

## Objectives:

You will be able to describe how to correctly assemble and inspect a SMAW outfit. You will also be able to show how to set the polarity and amperage on the machine.

## Instructions:

Read Chapter 10 and study Figures 10-1 through 10-10. Then, answer or complete the following questions.

_____ 1. Which of the following statements is *not* true for the output connections of an AC or DC arc welding machine?
A. They must always be tight.
B. Lead sockets or internal connections use a cam lock–style connector.
C. Wing nuts may be used to connect the welding leads.
D. Large slip-joint pliers may be used to tighten the hex nuts on the lead connections.
E. Coarse amperage ranges are selected on some machines by putting the electrode lead into various sockets.

_____ 2. Which of the following factors is a key consideration in determining the size (diameter) of the electrode and workpiece leads to be used?
A. The distance to the part being welded.
B. The open circuit voltage required.
C. The two-way distance between the job and the arc welding machine.
D. The polarity of the welding machine.
E. The type of lead connection being used.

3. *True or False?* The electrode holder and electrode should be on the worktable when the machine is started.

4. Name two sources for the 120V or 240V input power required for an AC or DC arc welding machine.

_____

_____

5. *True or False?* The diameters of the workpiece lead and the electrode lead should always be the same.

6. Name three ways, other than bolting, of connecting a workpiece lead to the base metal.

_____

_____

_____

7. *True or False?* For the best results, you should adjust the amperage or polarity while an arc is being drawn.

8. To change the fine amperage adjustment on arc welding machines, a welder would operate digital settings, a knob, or a(n) _____.

_____

_____ 9. What is a rectifier used for?
    A. To change DC current into AC current.
    B. To change AC current into DC current.
    C. To increase the current.
    D. To reduce the current.
    E. To reduce the voltage and increase the amperage.

10. What can be done or used to protect welding leads from damage if they must be temporarily laid across a busy aisle or high-traffic area in a shop or plant?

_____

_____

11. Provide two reasons why a wrapping of copper foil is used to connect the end of the electrode lead to the electrode holder.

_____

_____

12. List the eight steps required when inspecting a SMAW outfit.

_____
_____
_____
_____
_____
_____
_____
_____
_____
_____
_____
_____
_____
_____
_____
_____

13. To prevent damage to the arc welding machine, the electrode holder must always be hung on a(n) _____ _____ when it is not in the welder's hand.

_____

_____

Name: _____    Date: _____

Class: _____    Instructor: _____

Lesson Grade: _____    Instructor's Initials: _____

## Assigned Job 10-1
# Performing a Safety Inspection of a Shielded Metal Arc Welding Station

## Objective:

In this job, you will learn how to make a safety inspection of a SMAW station and report any unsafe conditions.

**Note**

Do not attempt this job until you have read all safety precautions, satisfactorily completed the *Shielded Metal Arc Welding (SMAW) Safety Test*, and been approved by your instructor.

1. Ask the instructor to assign you an SMAW station to inspect. Station number assigned: _____

**Caution**

Make certain that the machine power switch is in the *off* position before you begin your safety inspection.

2. Check the electrode and workpiece leads at the welding machine to make sure they are tight.

3. Check the electrode lead at the electrode holder to make certain it is tight. Check the workpiece lead at the table or at the base metal to make sure that it is tight.

4. Inspect the insulation on the outside of the electrode and workpiece leads. There should be no cuts or worn spots in the insulation.

5. Locate the place where the electrode holder is hung when it is not in the welder's hand. Make sure that this hook or holder is insulated from the welding electrical circuit.

6. There should be no dampness or water on the floor or welding table that could cause the welder to receive an electrical shock.

7. Locate the polarity switch (if used) and the amperage control. Check both to make certain that they move easily.

8. In the space provided, list any loose connections or unsafe conditions that you observed during your inspection.

_____

_____

_____

_____

_____

_____

_____

_____

# Lesson 11
# SMAW: Electrodes

## Objectives:

You will be able to identify carbon and low alloy electrodes. You will also be able to describe how to select the correct electrode, electrode diameter, polarity, and amperage for the weld being made.

## Instructions:

Read Chapter 11 and study Figures 11-1 through 11-11. Then, answer or complete the following questions.

_____ 1. The AWS A5.5 is a list of electrodes used to weld which type of metal?
   A. Copper and copper alloys.
   B. Carbon steels.
   C. Low alloy steels.
   D. Gray and ductile iron.
   E. Corrosion-resistant steels.

2. Electrodes are produced in lengths of 9" (229 mm), 12" (305 mm), ____ (____), and 18" (457 mm).

_____

_____ 3. Electrodes are produced in ____ different diameters.
   A. 10
   B. 9
   C. 8
   D. 7
   E. 16

_____ 4. The covering on an electrode has which of the following purposes?
   A. Add filler metal to the weld.
   B. Create a protective gas to protect the arc and metal.
   C. Strengthen the weld by adding alloying elements to the weld.
   D. Establish the polarity of the electrode.
   E. All of the above.

5. Some of the materials in the electrode covering are used to form a(n) ____ ____, which protects the weld from oxygen, nitrogen, dirt, and other airborne contaminants.

_____

6. List two things that are in the slag that forms on top of the welding bead during SMAW.

_____

_____    7.    _____ may be included in the electrode covering to increase the rate at which filler metal is deposited in the weld.
    A.   Flux
    B.   Iron powder
    C.   Metallic salts
    D.   Hydrogen
    E.   None of the above.

8.   What does the letter "E" mean in an AWS electrode designation?

_____

9.   *True or False?* The first two numbers in AWS E6013 and the first three numbers in AWS E12015 indicate the minimum tensile strength of the deposited metal.

_____    10.    The number 1 in AWS E6012 tells the welder that this electrode can be used in ____.
    A.   the flat welding position only
    B.   the flat and horizontal welding positions only
    C.   the flat welding position, for horizontal fillets only
    D.   all welding positions
    E.   vertical and overhead fillets only

_____    11.    The last two numbers in an AWS electrode number tell the polarity of the electrode. What polarity is used with an E7013?
    A.   DCEN.
    B.   DCEP.
    C.   DCSP.
    D.   AC.
    E.   DCEN or AC.

_____    12.    In the E7018-B1 electrode, the "B1" indicates that _____.
    A.   1/2% Mo has been added
    B.   1/2% Cr and 1/2% Mo have been added
    C.   1 1/4% Cr and 1% Mo have been added
    D.   iron powder has been added
    E.   3 1/4% Ni has been added

13.   In the following spaces, describe what each portion of the AWS E7018–B1 electrode label represents:

E: _____

70: _____ tensile strength.

1: _____ position(s).

18: _____ polarity.

–B1: contains _____

14.   *True or False?* The number 18 in the AWS 7018-B1 electrode tells the welder that it is a low-hydrogen electrode and the flux contains iron powder.

15.   Using the rule-of-thumb method described under the Electrode Amperage Requirements section of the textbook, what amperage range is suggested for a 5/32" diameter electrode? Show your work in the space provided.

_____

_____

Name _____

16. The suggested amperage range for a 5/32″ diameter E6013 electrode is _____–_____ amperes. Refer to Figure 11-9 in the text.

_____

_____   17.   To determine the correct diameter and type of electrode to use for a weld on a commercial bridge, the welder should look at ____.
A.   an electrode manufacturer's guide book
B.   a welder qualification designed by an engineer
C.   a welding procedure specification (WPS) for the weld being done
D.   past experience and knowledge
E.   Any of the above.

18. In the spaces provided, list any eight of the eleven factors that must be considered when choosing an electrode:

_____

_____

_____

_____

_____

_____

_____

_____

_____   19.   How wide is a normal stringer-type weld bead?
A.   As wide as the electrode diameter.
B.   2–3 times the diameter of the electrode.
C.   6 times as wide as the electrode diameter.
D.   4 times as wide as the electrode diameter.
E.   10 times as wide as the electrode diameter.

20. Briefly explain why electrodes must be kept dry.

_____

_____

_____

_____

# Lesson 12
# SMAW: Flat Welding Position

## Objectives:

You will be able to demonstrate how to make acceptable welds on butt, lap, inside corner, and T-joints in the flat welding position using SMAW.

## Instructions:

Read Chapter 12 and study Figures 12-1 through 12-22. Then, answer or complete the following questions.

1. *True or False?* Chipping goggles must always be worn when chipping the slag from an arc weld.

2. When welding a fillet weld on thin metal, the electrode should point more toward the _____ than toward the _____.

_____

_____ 3. Why is it possible to keep a thickly coated electrode in contact with the base metal without the arc stopping?
   A. A higher amperage is used.
   B. A low amperage is used.
   C. Thickly covered electrodes are moved faster, therefore the arc does not burn out.
   D. A high voltage is used.
   E. The thick covering does not conduct electricity.

_____ 4. At what point along the weld joint is the arc blow weakest?
   A. At the beginning.
   B. In the middle.
   C. At the end.
   D. Depends on the location of the workpiece lead.
   E. Depends on whether forehand or backhand welding is being performed.

_____ 5. What is the desired shape of the rear edge of the weld pool?
   A. Straight across.
   B. "V" shaped.
   C. Bullet-shaped.
   D. Rippled.
   E. Any of the above.

6. *True or False?* To restart the welding arc, strike the arc about 1/2"–1" (13 mm–25 mm) ahead of the old crater and move the welding arc back until it reaches the rear of the old crater. Then, begin to move the weld pool ahead again.

_____     7.  The distance between the end of the electrode and the surface of the base metal is
                called the ____.
                A.  the electrode distance
                B.  the arc distance
                C.  the arc length
                D.  the resistance length
                E.  the electrode length

8.  *True or False?* Welding while standing on a floor that is damp (wet) or made of metal can cause
    an electrical shock.

9.  "Reading a bead" can provide information about the ____, ____ ____, and travel speed used
    when making a weld.

_____

_____

10. *True or False?* When arc blow occurs, the filler metal has a tendency to blow toward the ends of
    the weld joint.

_____     11.  When making a fillet weld on a T-joint or inside corner joint, the electrode is held at a
                ____ work angle.
                A.  30°
                B.  45°
                C.  75°
                D.  90°

12. The electrode may be held at a(n) ____° travel angle while making any weld bead with SMAW.

_____

13. What type weld bead is being made in the following figure?

_____

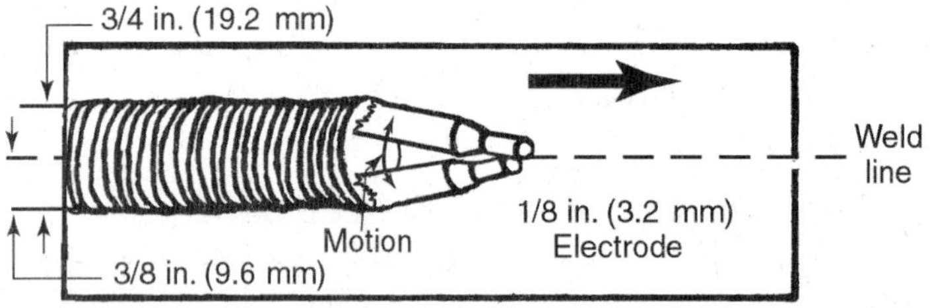

14. How wide should a stringer bead be?
    A.  The diameter of the electrode.
    B.  Two to three times the electrode diameter.
    C.  Three to four times the electrode diameter.
    D.  Five times the electrode diameter.
    E.  Six to eight times the electrode diameter.

15. List two methods of striking the arc to begin a SMAW.

_____

_____

Name _____

_____   16.   How high should the weld bead be above the base metal?
        A.   1/16 the electrode diameter.
        B.   1/8 the electrode diameter.
        C.   1/8 the weld bead width.
        D.   1/2 the weld bead width.
        E.   Equal to the electrode diameter.

17.   What will cause the welding arc to go out or stop? List two causes:

_____

_____

_____   18.   When it is necessary to release an electrode that is stuck to the base metal, which of
        the following activities is not done?
        A.   Turn the amperage down.
        B.   Keep your arc welding helmet down and release the electrode from the electrode
            holder.
        C.   Lift your helmet after the electrode is released from the electrode holder and place
            the electrode holder on an insulated hook.
        D.   Hold the electrode near the base metal and bend it back and forth until it is free of
            the base metal.

19.   When arc welding, what must be done to prevent dirt and slag from one weld bead from mixing
    with weld beads that are created later?

_____

_____

_____   20.   The correct welding arc length is about ____.
        A.   1/8″ (3.2 mm)
        B.   3/16″ (4.8 mm)
        C.   5/32″ (4.0 mm)
        D.   less than 1/8″ (3.2 mm)
        E.   equal to the diameter of the electrode

21.   If parts cleaned with ____ ____ are not thoroughly dried before they are welded, a poisonous
    gas called phosgene can be created.

_____

_____   22.   To counteract forward arc blow, place the workpiece lead ____.
        A.   near the start of the weld joint
        B.   near the middle of the weld joint
        C.   near the end of the weld joint

*For Questions 23–25, refer to the following illustration*

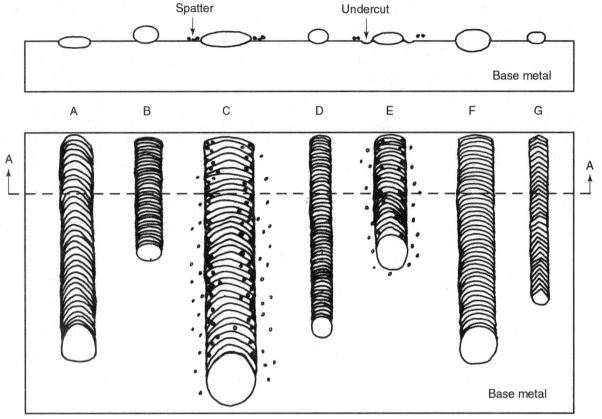

American Welding Society (AWS)

_____ 23. How would you correct weld bead E?
   A. Increase the amperage.
   B. Increase the travel speed.
   C. Decrease the amperage.
   D. Decrease the arc length.
   E. Decrease the travel speed.

24. Which weld bead was made with too much amperage?

_____

25. Which weld bead was made with a travel speed that was too fast?

_____

Name: _____ Date: _____

Class: _____ Instructor: _____

Lesson Grade: _____ Instructor's Initials: _____

# Running a Weld Bead in the Flat Welding Position

## Objectives:

In this job, you will learn to make satisfactory stringer beads using the SMAW process. You will also learn to read a weld bead.

> **Note**
>
> Do not attempt this job until you have read all safety precautions, satisfactorily completed the *Shielded Metal Arc Welding (SMAW) Safety Test*, and been approved by your instructor.

1. Obtain three pieces of mild steel that measures 1/4″ × 5″ × 6″ (6.4 mm × 125 mm × 150 mm). Also, obtain three each of the following electrodes:
   - 1/8″ (3.2 mm) diameter E6010.
   - 5/32″ (4 mm) diameter E6012.
   - 1/8″ (3.2 mm) diameter E6013.
   - 5/32″ (4 mm) diameter E7018 (or other iron powder electrode).

2. Fill in the amperage range and polarity for each of the electrodes. Refer to Figure 11-9 in the text for the amperage ranges and Figure 11-6 for the suggested polarity.
   A. Amperage range and polarity:

| Diameter | Designation | Amperage Range | | Polarity |
|:---:|:---:|:---:|:---:|:---:|
| | | **Min.** | **Max.** | |
| 1/8″ | E6010 | | | |
| 5/32″ | E6012 | | | |
| 1/8″ | E6013 | | | |
| 5/32″ | E7018 | | | |

   B. How wide should a stringer bead be for a 1/8″ (3.2 mm) diameter electrode? Show your work.

   _____

   _____

   C. How wide should a stringer bead be for a 5/32″ (4.0 mm) diameter electrode? Show your work.

   _____

   _____

D. What is the maximum width for a weave bead made with a 1/8″ (3.2 mm) diameter electrode? Show your work.

_____

E. What is the maximum width for a weave bead made with a 5/32″ (4.0 mm) diameter electrode? Show your work.

_____

3. Make a safety inspection of your welding station.

4. Place the 1/4″ (6.4 mm) steel plate on the work table.

5. Place the E6010 electrode into the electrode holder and hang the holder on the insulated hook in the booth.

6. Set the amperage on the welding machine for the middle of the correct amperage range and set the correct polarity on the machine.

7. See Figure 12-9 in the text for suggested angles to use when running weld beads with SMAW.

8. Make a few practice stringer beads to set the amperage for the best results.

9. Use a chipping hammer and wire brush to clean each weld bead before the next weld bead is made.

**Caution**
Always wear approved chipping goggles when chipping or wire brushing a weld bead.

10. Draw lines on the metal with a soapstone to help keep the weld beads straight. Make two stringer beads along the 6″ (150 mm) length of the practice plate. These weld beads should be about 1/2″ (13 mm) apart. Watch the weld bead at all times with every electrode used. If the weld bead does not look right, stop immediately. Read the weld bead, make required changes, and restart the weld bead.

11. Make two stringer beads with each of the other three electrode types and diameters. Use the other side of the base metal, if necessary. The drag welding method may be used with the E7018 electrode.

12. Make a few practice weave beads.

13. Make one weave bead with each of the four electrode types and diameters for a grade.

14. Use chalk or soapstone to identify the electrode used for each weld bead.

## Inspection:

Each weld bead should be straight and even in width and height. The ripples must be even. Restarts should be smooth, with no low spots or unfilled areas. Stringer beads and weave beads should be of the correct width. Complete penetration is not required.

## Assigned Job 12-2
# Welding a Lap Joint in the Flat Welding Position

## Objective:

In this job, you will learn to make an acceptable fillet weld on a lap joint in the flat welding position.

> **Note**
> Do not attempt this job until you have read all safety precautions, satisfactorily completed the *Shielded Metal Arc Welding (SMAW) Safety Test*, and been approved by your instructor.

1. You will need four pieces of low-carbon steel that measure 1/4″ × 1 1/2″ × 6″ (6.4 mm × 40 mm × 150 mm). You will also need four E6010 electrodes.

2. Answer the following questions:
   A. Determine the correct diameter of E6010 electrode to use for welding 1/4″ thick low-carbon steel. Refer to Figure 11-9 in the text. _____
   B. What amperage range is suggested for the E6010 electrode? Refer to Figure 11-9 in the text. _____–_____ amperes.
   C. What polarity is suggested for the E6010 electrode? Refer to Figure 11-6 in the text. _____

3. Set the machine amperage for the middle of the suggested amperage range. Also set the correct polarity.

4. Make a practice weld to set the machine for the best welding results.

5. Arrange two pieces of metal as shown in the following illustration. Tack weld the weldment in three places about 3″ (75 mm) apart. Flip the weldment over and tack weld the other side the same way.

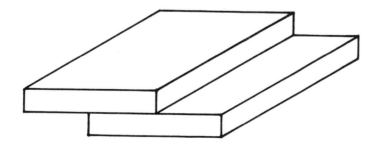

6. After the pieces have been tack welded, arrange the weldment as shown in the following illustration. The weld face must be horizontal when welding in the flat welding position.

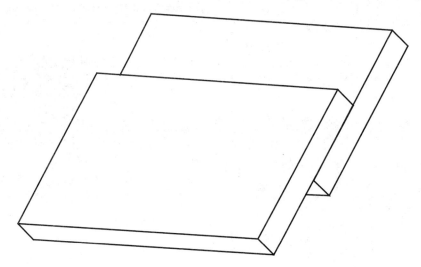

7. The electrode angles for welding a fillet weld on a lap joint in the flat welding position are shown in Figure 12-17 in the text.

8. Make two weldments like the one shown in step 5. Remember to read the weld at all times and change the amperage, welding speed, electrode angles, and arc length as necessary to make good fillet welds.

9. Clean the weld beads with a chipping hammer and wire brush.

**Caution**
Always wear approved chipping goggles when chipping or wire brushing a weld bead.

## Inspection:

The weld beads must be convex and straight, with evenly spaced ripples. Both legs of the fillet weld should be 1/4″ (6 mm). There should be no evidence of undercut or overlap. The weld beads should be well-fused into the base metal.

Name: _____    Date: _____

Class: _____    Instructor: _____

Lesson Grade: _____    Instructor's Initials: _____

## Assigned Job 12-3
# Welding a T-Joint in the Flat Welding Position

### Objective:

In this job, you will learn to make an acceptable fillet weld on a T-joint in the flat welding position.

> **Note**
>
> Do not attempt this job until you have read all safety precautions, satisfactorily completed the *Shielded Metal Arc Welding (SMAW) Safety Test*, and been approved by your instructor.

1. Obtain four pieces of low-carbon steel that measure 1/4″ × 1 1/2″ × 6″ (6.4 mm × 40 mm × 150 mm) and two pieces that measure 1/4″ × 3″ × 6″ (6.4 mm × 75 mm × 150 mm). You will also need six 1/8″ diameter E6012 electrodes.

2. Answer the following questions:
   A. What amperage range is suggested for the E6012 electrode. Refer to Figure 11-9 in the text. _____–_____ amperes.
   B. What is the correct polarity for this electrode? Refer to Figure 11-6 in the text. _____

3. Set the machine amperage for the middle of the recommended range. Set the polarity also.

4. Make a few practice welds to adjust the machine for the best weld beads.

5. Arrange one of the large pieces and one of the small pieces as shown in the following illustration. Tack the weldment in three places about 3″ (75 mm) apart. The tack welds may be made with the welds in the horizontal welding position.

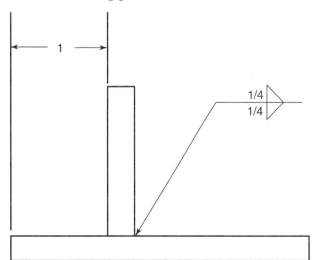

| in. | mm |
|-----|------|
| 1/4 | 6.4 |
| 1 | 25.4 |

6.  After cleaning the tack welds, arrange the pieces so that the welding can be done in the flat welding position.

7.  See Figure 12-17 in the text for the correct electrode angles to use when welding a fillet weld on a lap joint in the flat welding position.

8.  Make the fillet welds indicated in the drawing for step 5. Weld in the flat position.

9.  Clean each weld bead by chipping and wire brushing.

**Caution**
Always wear approved chipping goggles when chipping and wire brushing.

10.  Place a second 1 1/2″ (40 mm) wide piece as shown in the following illustration. Tack weld the pieces only on the weld side of the joint.

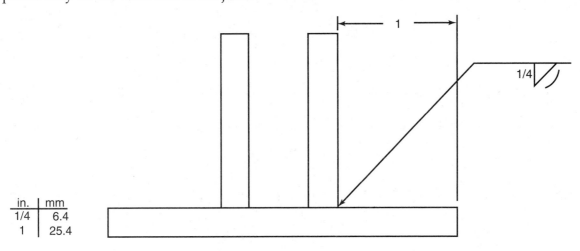

| in. | mm |
|-----|------|
| 1/4 | 6.4 |
| 1 | 25.4 |

11.  After cleaning the tack welds, make the third weld on the weldment as indicated in the drawing for step 10.

12.  Repeat steps 5–11, using the other three pieces of metal.

## Inspection:

The welds must be made according to the AWS welding symbol. The weld beads must be straight, convex in shape, and have uniformly and evenly spaced ripples. All restarts should be made without low spots.

Name: _____     Date: _____

Class: _____     Instructor: _____

Lesson Grade: _____     Instructor's Initials: _____

## Assigned Job 12-4
# Welding a V-Groove Outside Corner Joint in the Flat Welding Position

### Objective:

You will learn to make an acceptable V-groove weld on an outside corner joint in the flat welding position. This weld is similar in many respects to a V-groove butt joint.

---

**Note**

Do not attempt this job until you have read all safety precautions, satisfactorily completed the *Shielded Metal Arc Welding (SMAW) Safety Test*, and been approved by your instructor.

---

1. Obtain four pieces of low-carbon steel that measure 1/4″ × 1 1/2″ × 6″ (3.2 mm × 40 mm × 150 mm). Also obtain six 1/8″ diameter E6010 electrodes.

2. Answer the following questions before welding:
   A. What amperage range is suggested for a 1/8″ (3.2 mm) diameter E6010 electrode? Refer to Figure 11-9 in the text. _____–_____ amperes.
   B. What polarity is suggested for this electrode? Refer to Figure 11-6 in the text. _____

3. Set the machine amperage at the middle of the suggested amperage range. Also, set the correct polarity.

4. Make a few practice weld beads. Read the weld beads and set the machine for the best results.

5. Arrange the pieces as shown in the following illustration. Tack weld each joint in three places. Make certain that the root opening is correct in each joint.

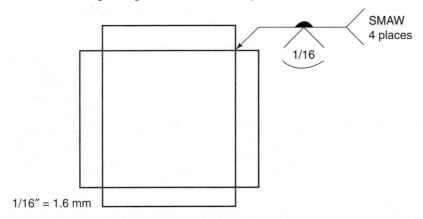

6. See Figures 12-9 and 12-10 in the text for the correct electrode angle to use when making a weld on a butt joint in the flat welding position.

7. Weld each joint as indicated in the drawing for step 5. As the weld is made, a keyhole must be present at all times. The keyhole ensures that the weld will have 100% penetration to the root of the weld.

8. Clean each bead by chipping and wire brushing.

**Caution**
Always wear chipping goggles when chipping or wire brushing.

## Inspection:

The welds in these V-groove butt joints must be made as shown in the AWS welding symbol. The weld beads must be straight, with a convex shape and evenly spaced ripples. Each weld must have evidence of 100% penetration on the root side. All restarts should be made without low spots.

## Lesson 13
# SMAW: Horizontal, Vertical, and Overhead Welding Positions

## Objectives:

You will be able to demonstrate how to make acceptable out-of-position SMAW welds on butt, lap, inside corner, outside corner, and T-joints.

## Instructions:

Read Chapter 13 and study Figures 13-1 through 13-14. Then, answer or complete the following questions.

_____    1. It is best to use an electrode with a slag that cools quickly when welding out of position. This helps to keep the molten metal from sagging. Which of the electrodes listed below is not suitable for welding out of position?
   A. E6013.
   B. E9011.
   C. E6010.
   D. E7028.
   E. E7015.

2. When welding downhill, slag should not be allowed to enter the weld pool. Why?

   _____

3. *True or False?* Out-of-position welds may be made with many weld passes.

4. A travel angle of ____°–____° is used when welding vertically, regardless of the joint type.

5. *True or False?* The electrode angles used while welding vertically are the same as those used when welding in the flat welding position.

6. If a drag angle of 40° is used when welding a fillet weld in the horizontal position, what will probably happen to the weld bead?

   _____

7. Name five welding clothing or equipment items that should be worn by a welder performing out-of-position welds.

   _____

8. *True or False?* Electrodes may be placed into the electrode holder at various angles, but should never be bent.

_____    9.  How do the electrode angles used for a fillet weld in the horizontal position differ from the electrode angles used for a fillet weld in the flat welding position?
    A.   The electrode is tilted forward more.
    B.   The angle from the electrode to the base metal is greater.
    C.   The electrode is tilted backward farther.
    D.   The angle from the electrode to the base metal is smaller.
    E.   The angles are the same in both positions.

_____    10.  A sharp, pointed shape at the rear of the weld pool indicates _____.
    A.   the amperage is too high
    B.   the amperage is too low
    C.   the welding speed is too fast
    D.   the welding speed is too slow
    E.   the bead width is too small

_____    11.  If the toes of a fillet weld are not well-fused, _____ the amperage or _____ the welding speed.
    A.   increase, increase
    B.   decrease, decrease
    C.   increase, decrease
    D.   decrease, increase

_____    12.  To prevent the molten metal from sagging while making a horizontal butt weld, the electrode is pointed upward from the weld axis about _____.
    A.   5°–10°
    B.   10°–20°
    C.   30°–45°
    D.   5°–15°
    E.   15°–45°

13.  What type of weld defect is caused when slag is trapped during a multiple-pass weld?

_____

14.  Describe the whip motion.

_____

_____

15.  What type of motion, used while welding a horizontal lap joint, helps to prevent the molten metal from sagging?

_____

16.  When welding in the overhead welding position, a(n) _____ weld pool should be maintained to keep the metal from sagging.

_____

17.  List three reasons for using a smaller-diameter electrode when welding out of position.

_____

_____

_____

Name _____

18.  List three SMAW electrodes that have fast-setting slag deposits.

_____

_____   19.   Which of the following is *not* an advantage of using a smaller electrode?
   A.   A lower amperage can be used.
   B.   A longer arc length can be used.
   C.   The molten metal is easier to control.
   D.   A smaller weld is created.

_____   20.   What does a keyhole-shaped weld pool ensure?
   A.   Full penetration on the root pass of a butt weld.
   B.   Full penetration on the root pass of a fillet weld.
   C.   Full penetration on the cover pass of a butt weld.
   D.   Full penetration on the cover pass of a fillet weld.

Name: _____   Date: _____

Class: _____   Instructor: _____

Lesson Grade: _____   Instructor's Initials: _____

## Assigned Job 13-1
# Welding a Lap Joint in the Horizontal Welding Position

## Objective:

In this job, you will learn to make an acceptable fillet weld on a lap joint in the horizontal welding position.

> **Note**
> Do not attempt this job until you have read all safety precautions, satisfactorily completed the *Shielded Metal Arc Welding (SMAW) Safety Test*, and been approved by your instructor.

1. You will need four pieces of low-carbon steel that measure 1/4″ × 1 1/2″ × 6″ (6.4 mm × 40 mm × 150 mm) and one piece that measures 1/4″ × 3″ × 6″ (6.4 mm × 75 mm × 150 mm). You will also need six 1/8″ (3.2 mm) diameter E6011 electrodes.

2. Answer the following questions by referring to Figures 11-6 and 11-9 in the text:
   A. What amperage range is suggested for the E6011 electrode? ____–____ amperes.
   B. What DC polarity is used with an E6011 electrode? _____

3. Set the welding machine amperage for the middle of the suggested amperage range. Set the correct DC polarity.

4. Place two 1 1/2″ (40 mm) pieces together to form a lap joint with a 3/4″ (19 mm) overlap. Tack weld these pieces in three places on each side. The tack welding may be done in the flat welding position.

5. Make a few practice welds to set the welding machine for the best welding results. See Figure 13-6 in the text for the correct electrode angle to use while making a fillet weld in the horizontal welding position.

6. Clean the weld beads with a chipping hammer and a wire brush. Read the weld beads and make whatever changes are required.

> **Caution**
> Always wear approved chipping goggles when chipping or wire brushing the slag from a weld bead.

7. Arrange the remaining three pieces as shown in the following drawing.

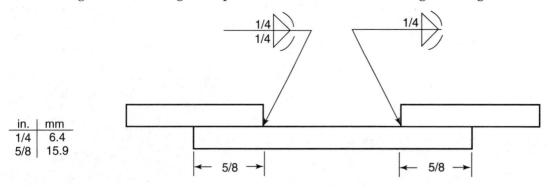

| in. | mm |
| --- | --- |
| 1/4 | 6.4 |
| 5/8 | 15.9 |

8. Tack weld each joint in three places.

9. Make the fillet welds shown by the AWS symbols at each of the lap joints.

10. Clean each weld bead with a chipping hammer and wire brush.

## Inspection:

All weld beads should be straight and have an even width. The weld beads must conform in size and shape to the AWS welding symbol. No undercut or overlap should be present. The toes of each fillet should show good fusion with the base metal. The ripples should be even, bullet-shaped, and have no low spots.

## Assigned Job 13-2
# Welding a T-Joint in the Horizontal Welding Position

## Objective:

In this job, you will learn to make an acceptable fillet weld on a T-joint in the horizontal welding position.

> **Note**
> Do not attempt this job until you have read all safety precautions, satisfactorily completed the *Shielded Metal Arc Welding (SMAW) Safety Test*, and been approved by your instructor.

1. You will need four pieces of low-carbon steel that measure 1/4″ × 1 1/2″ × 6″ (6.4 mm × 40 mm × 150 mm) and one piece that measures 1/4″ × 3″ × 6″ (6.4 mm × 75 mm × 150 mm). You will also need six to eight 1/8″ (3.2 mm) diameter E6012 electrodes.

2. Answer the following questions by referring to Figures 11-6 and 11-9 in the text:
   A. What is the suggested amperage range for the E6012 electrode? ____–____ amperes.
   B. What DC polarity is used for an E6012 electrode? _____

3. Set the welding machine amperage to the middle of the amperage range. Set the correct polarity.

4. Place one of the 1 1/2″ (40 mm) pieces on top of another 1 1/2″ (40 mm) piece to form a T-joint. Tack weld the T-joint in three places on each side.

5. Make two practice welds to set the welding machine. See Figure 13-6 in the text for the correct electrode angles to use while welding a T-joint in the horizontal welding position.

6. Clean the weld beads with a chipping hammer and a wire brush.

> **Caution**
> Always wear approved chipping goggles when chipping or wire brushing the slag from a weld bead.

7. Arrange the 3″ (75 mm) piece and one of the remaining 1 1/2″ (40 mm) pieces as shown in the following drawing.

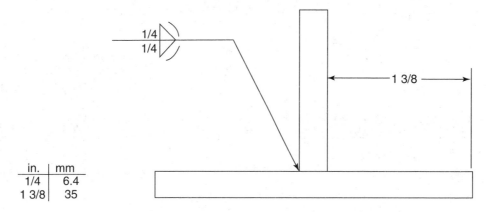

| in. | mm |
| --- | --- |
| 1/4 | 6.4 |
| 1 3/8 | 35 |

8. Tack weld the joints three times on each side.

9. Make the fillet welds as shown by the AWS weld symbol in step 7.

10. Place the last piece of 1 1/2″ (40 mm) metal onto the weldment from step 9, as shown in the following drawing. Tack weld and weld the joint according to the welding symbol and drawing. Place the weldment in a weld positioner if needed to weld in the horizontal position.

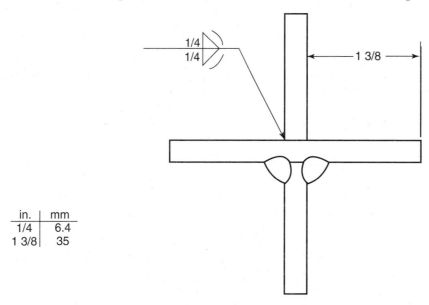

| in. | mm |
| --- | --- |
| 1/4 | 6.4 |
| 1 3/8 | 35 |

11. Clean the welds with a chipping hammer and wire brush.

## Inspection:

All the weld beads must be straight and even in width. They must conform in size and shape to the AWS weld symbol. The toes of the fillet must show good fusion with the base metal. The ripples should be evenly spaced, bullet-shaped, and have no low spots.

## Assigned Job 13-3
# Welding a V-Groove Butt Joint in the Horizontal Welding Position

## Objective:
In this job, you will learn to make an acceptable V-groove weld in the horizontal welding position.

**Note**
Do not attempt this job until you have read all safety precautions, satisfactorily completed the *Shielded Metal Arc Welding (SMAW) Safety Test*, and been approved by your instructor.

1.  You will need five pieces of low-carbon steel that measure 1/4″ × 1 1/2″ × 6″ (6.4 mm × 40 mm × 150 mm). You will also need six to eight 1/8″ (3.2 mm) diameter E6010 electrodes.

2.  Answer the following questions by referring to Figures 11-6 and 11-9 in the text:
    A.  What amperage range is suggested for the E6010 electrode? ____–____ amperes.
    B.  What DC polarity is used with an E6010 electrode? _____

3.  Set the welding machine amperage for the middle of the suggested amperage range. Set the correct DC polarity.

4.  Use two pieces for practice. Cut or grind one edge of each piece to form a 35° bevel.

**Caution**
Wear approved grinding goggles whenever you are grinding.

5.  Place the beveled edges of the two pieces together to form a single V-groove joint. Tack weld these pieces in three places. The tack welding may be done in the flat welding position.

6.  Place the weldment in a welding positioner. Make practice welds to set the welding machine for the best welding results. See Figure 13-9 in the text for the correct electrode angles to use while welding a V-groove butt weld in the horizontal welding position. A keyhole should be seen at all times as this weld is made. This will ensure that 100% penetration is occurring.

7.  Clean the weld bead with a chipping hammer and a wire brush. Read the weld beads and make whatever changes are required.

**Caution**
Always wear approved chipping goggles when chipping or wire brushing the slag from a weld bead.

8. Cut or grind the edges of the remaining pieces as required. Arrange the pieces as shown in the following drawing.

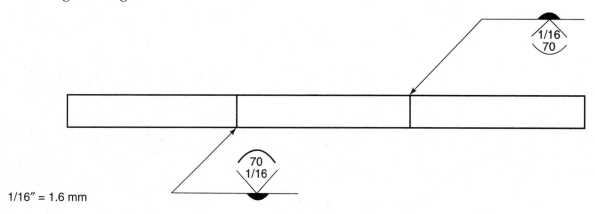

1/16″ = 1.6 mm

9. Tack weld each joint in three places. This may be done in the flat welding position.

10. Place the weldment into a welding positioner so the welds can be made in the horizontal welding position.

11. Make the V-groove butt welds shown by the AWS symbols.

12. Clean each weld bead with a chipping hammer and wire brush.

## Inspection:

All weld beads should be straight and have an even width. The weld beads must conform in size and shape to the AWS welding symbol. No undercut or overlap should be present. The toes of each weld should show good fusion with the base metal. The ripples should be even, bullet-shaped, and have no low spots. There must be evidence of 100% penetration on the root side of each weld.

## Assigned Job 13-4
# Welding a Lap Joint in the Vertical Position

### Objective:

In this job, you will learn to make an acceptable fillet weld on a lap joint in the vertical welding position.

**Note**

Do not attempt this job until you have read all safety precautions, satisfactorily completed the *Shielded Metal Arc Welding (SMAW) Safety Test*, and been approved by your instructor.

1. You will need five pieces of low-carbon steel that measure 1/8″ × 1 1/2″ × 6″ (3.2 mm × 40 mm × 150 mm). You will also need six to eight 3/32″ (2.4 mm) E6012 electrodes.

2. Answer the following questions by referring to Figures 11-6 and 11-9 in the text:
   A. What amperage range is suggested for the E6012 electrode? _____–_____ amperes.
   B. What DC polarity is used with an E6012 electrode? _____

3. Set the welding machine amperage for the middle of the suggested amperage range. Set the correct DC polarity.

4. Place two 1 1/2″ (40 mm) pieces together to form a lap joint with a 3/4″ (19 mm) overlap. Tack weld these pieces in three places on each side. The tack welding may be done in the horizontal welding position.

5. Place the weldment in a welding positioner. Make two practice welds in the vertical welding position to set the welding machine for the best results. The weldment may be moved as required to make all welds in the vertical welding position. The correct electrode angles to use while welding a fillet in the vertical welding position are the same as those used in the flat or horizontal welding positions. See Figure 13-6 in the textbook.

6. Clean the weld beads with a chipping hammer and a wire brush. Read the weld beads and make whatever changes are required.

**Caution**

Always wear approved chipping goggles when chipping or wire brushing the slag from a weld bead.

7.  Arrange the remaining three pieces as shown in the drawing.

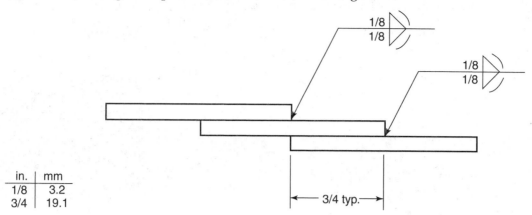

| in. | mm |
|-----|------|
| 1/8 | 3.2 |
| 3/4 | 19.1 |

8.  Tack weld each joint in three places. The tack welding may be done in the horizontal welding position.

9.  Place the weldment into a welding positioner, so that the welds can be made in the vertical welding position. The weldment may be moved so that each weld is made in the vertical welding position.

10. Make the fillet welds shown by the AWS symbols on each of the lap joints.

11. Clean each weld bead with a chipping hammer and wire brush.

## Inspection:

All weld beads should be straight and have an even width. The weld beads must conform in size, shape, and location to the AWS welding symbol. No undercut or overlap should be present. The toes of each fillet should show good fusion with the base metal. The ripples should be even, bullet-shaped, and have no low spots.

## Assigned Job 13-5
# Welding a T-Joint in the Vertical Welding Position

## Objective:

In this job, you will learn to make an acceptable fillet weld on a T-joint in the vertical welding position.

**Note**

Do not attempt this job until you have read all safety precautions, satisfactorily completed the *Shielded Metal Arc Welding (SMAW) Safety Test*, and been approved by your instructor.

1. You will need two pieces of low-carbon steel that measure 1/4″ × 1 1/2″ × 6″ (6.4 mm × 40 mm × 150 mm) and two pieces that measure 1/4″ × 3″ × 6″ (6.4 mm × 75 mm × 150 mm). You will also need six to eight 1/8″ (3.2 mm) diameter E6013 electrodes.

2. Answer the following questions by referring to Figures 11-6 and 11-9 in the text:
   A. What amperage range is suggested for the E6013 electrode? _____–_____ amperes.
   B. What DC polarity is used with an E6013 electrode? _____

3. Set the welding machine amperage for the middle of the suggested amperage range. Set the correct DC polarity.

4. Place one 1 1/2″ (40 mm) piece on a 3″ (75 mm) piece to form a T-joint. Tack weld these pieces in three places on each side. The tack welding may be done in the flat welding position.

5. Place the weldment in a welding positioner. Make two practice welds in the vertical welding position to set the welding machine for the best welding results. The correct electrode angles to use while welding a fillet in the vertical welding position are the same as those used in the flat or horizontal welding positions. See Figure 13-6 in the text.

6. Clean the weld beads with a chipping hammer and a wire brush. Read the weld beads and make whatever changes are required.

**Caution**

Always wear approved chipping goggles when chipping or wire brushing the slag from a weld bead.

7.  Arrange the other two pieces as shown in the following drawing.

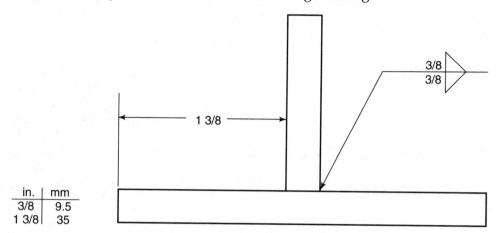

| in. | mm |
|-----|-----|
| 3/8 | 9.5 |
| 1 3/8 | 35 |

8.  Tack weld each joint in three places. The tack welding may be done in the flat welding position.

9.  Position the weldment in the welding positioner, so that each weld can be made in the vertical welding position.

10. Make the fillet welds shown by the drawing and AWS symbols in step 7.

11. Clean each weld bead with a chipping hammer and wire brush.

## Inspection:

All weld beads should be straight and have an even width. The weld beads must conform in size and shape to the AWS welding symbol. No undercut or overlap should be present. The toes of each weld should show good fusion with the base metal. The ripples should be even, bullet-shaped, and have no low spots.

## Assigned Job 13-6
# Welding a V-Groove Butt Joint in the Vertical Welding Position

## Objective:

In this job, you will learn to make an acceptable V-groove weld on a butt joint in the vertical welding position.

---

**Note**

Do not attempt this job until you have read all safety precautions, satisfactorily completed the *Shielded Metal Arc Welding (SMAW) Safety Test*, and been approved by your instructor.

---

1. You will need five pieces of low-carbon steel that measure $1/4'' \times 1\ 1/2'' \times 6''$ (6.4 mm × 40 mm × 150 mm). You will also need six $1/8''$ (3.2 mm) diameter E6013 electrodes.

2. Answer the following questions by referring to Figures 11-6 and 11-9 in the text:
   A. What amperage range is suggested for the E6013 electrode? ____–____ amperes.
   B. What DC polarity is used with an E6013 electrode? _____

3. Set the welding machine amperage for the middle of the suggested amperage range. Set the correct DC polarity.

4. Flame cut or grind one edge on each piece to create a 35° bevel. One piece requires two edges to be cut or ground. Study the drawing in step 8 to determine which edge to prepare.

---

**Caution**

Wear approved grinding goggles whenever you are grinding.

---

5. Place the two pieces together to form a V-groove butt joint. Tack weld the joint in three places. The tack welding may be done in the flat welding position.

6. Place the weldment in a welding positioner. Make a practice weld in the vertical welding position to set the welding machine for the best welding results. The correct electrode angles to use while welding a V-groove butt joint in the vertical welding position are shown in Figure 13-10 and 13-11 in the text.

7. Clean the weld beads with a chipping hammer and a wire brush. Read the weld beads and make whatever changes are required.

---

**Caution**

Always wear approved chipping goggles when chipping or wire brushing the slag from a weld bead.

---

8. Arrange the remaining three pieces as shown in the following drawing:

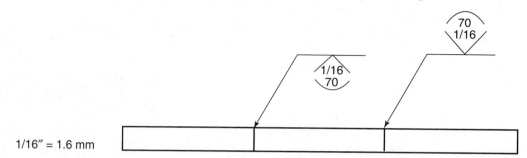

1/16″ = 1.6 mm

9. Tack weld each joint in three places. The tack welding should be done in the weld groove. The joint may be tack welded in the flat welding position.

10. Place the weldment into a welding positioner, so that the welds can be made in the vertical welding position.

11. Make the V-groove butt welds as shown by the drawing and AWS symbols.

12. Clean each weld bead with a chipping hammer and wire brush.

## Inspection:

All weld beads should be straight and have an even width. There should be full penetration. The weld beads must be the right size and shape, and in the right place according to the AWS welding symbols. No undercut or overlap should be present. The toes of each fillet should show good fusion with the base metal. The ripples should be even, bullet-shaped, and have no low spots.

## Assigned Job 13-7
# Welding a Lap Joint in the Overhead Welding Position

## Objective:

In this job, you will learn to make an acceptable fillet weld on a lap joint in the overhead welding position.

**Note**
Do not attempt this job until you have read all safety precautions, satisfactorily completed the *Shielded Metal Arc Welding (SMAW) Safety Test*, and been approved by your instructor.

1. You will need five pieces of low-carbon steel that measure 1/8″ × 1 1/2″ × 6″ (3.2 mm × 40 mm × 150 mm). You will also need six 3/32″ (2.4 mm) diameter E6012 electrodes.

2. Answer the following questions by referring to Figures 11-6 and 11-9 in the text:
   A. What amperage range is suggested for the E6012 electrode? _____–_____ amperes.
   B. What DC polarity is used with an E6012 electrode? _____

3. Set the welding machine amperage at the lower end of the suggested amperage range. Set the correct DC polarity.

4. Place two pieces together with a 3/4″ (19 mm) overlap to form a lap joint. Tack weld each joint in three places. The tack welding may be done in the flat welding position.

5. Place the weldment in a welding positioner. Make two practice welds in the overhead welding position to set the welding machine for the best results. The correct electrode angles to use while welding a fillet weld on a lap joint in the overhead welding position are the same as those used in the flat welding position.

6. Clean the weld beads with a chipping hammer and a wire brush. Read the weld beads and make whatever changes are required.

**Caution**
Always wear approved chipping goggles when chipping or wire brushing the slag from a weld bead.

7. Arrange two of the remaining pieces as shown in the following drawing.

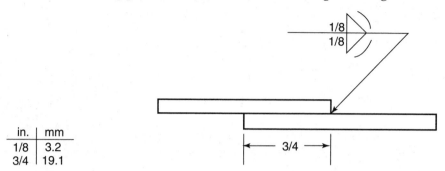

| in. | mm |
|-----|------|
| 1/8 | 3.2 |
| 3/4 | 19.1 |

8. Tack weld each joint in three places. The tack welding may be done in the flat welding position.

9. Place the weldment into a welding positioner so that the welds can be made in the overhead welding position. The weldment is turned as required to make all welds in the overhead welding position.

10. Make the fillet welds as shown by the drawing and AWS symbol shown in step 7.

11. Clean each weld bead with a chipping hammer and wire brush.

12. Place a third piece on the weldment as shown in the following drawing. Tack weld it. Make the last weld in the overhead welding position, as shown by the AWS welding symbol in the drawing.

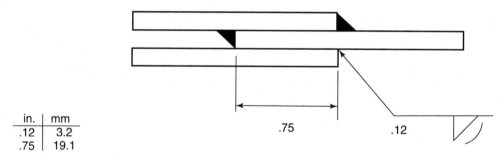

| in. | mm |
|-----|------|
| .12 | 3.2 |
| .75 | 19.1 |

## Inspection:

All weld beads should be straight and have an even width. The weld beads must be the right size and shape, and in the right place according to the AWS welding symbols. No undercut or overlap should be present. The toes of each fillet should show good fusion with the base metal. The ripples should be even, bullet-shaped, and have no low spots.

## Assigned Job 13-8
# Welding an Inside Corner Joint in the Overhead Welding Position

## Objective:

In this job, you will learn to make an acceptable fillet weld on an inside corner in the overhead welding position.

> **Note**
> Do not attempt this job until you have read all safety precautions, satisfactorily completed the *Shielded Metal Arc Welding (SMAW) Safety Test*, and been approved by your instructor.

1. You will need four pieces of low-carbon steel that measure 1/4″ × 1 1/2″ × 6″ (3.2 mm × 40 mm × 150 mm), and one that measures 1/4″ × 3″ × 6″ (3.2 mm × 75 mm × 150 mm). You will also need six to eight 1/8″ (3.2 mm) diameter E6013 electrodes.

2. Answer the following questions by referring to Figures 11-6 and 11-9 in the text:
   A. What amperage range is suggested for the E6013 electrode? _____–_____ amperes.
   B. What DC polarity is used with an E6013 electrode? _____

3. Set the welding machine amperage at the lower end of the suggested amperage range. Set the correct DC polarity.

4. Place two pieces of 1 1/2″ (40 mm) metal together to form a T-joint. Tack weld the joint in three places on each side. The tack welding may be done in the flat welding position.

5. Place the weldment in a welding positioner. Make two practice welds in the overhead welding position to set the welding machine for the best results. The correct electrode angles to use while welding a fillet weld on a T-joint in the overhead welding position are the same as those used in the flat welding position.

6. Clean the weld beads with a chipping hammer and a wire brush. Read the weld beads and make whatever changes are required.

> **Caution**
> Always wear approved chipping goggles when chipping or wire brushing the slag from a weld bead.

7. Arrange the three remaining pieces as shown in the following drawing.

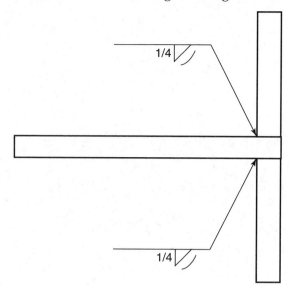

1/4″ = 6.4 mm

8. Tack weld each joint in three places. The tack welding may be done in the horizontal welding position.

9. Place the weldment into a welding positioner so that the welds can be made in the overhead welding position. The weldment is turned as required to make all welds in the overhead welding position.

10. Make the fillet welds as shown by the drawing and AWS symbols in step 7.

11. Clean each weld bead with a chipping hammer and wire brush.

## Inspection:

All weld beads should be straight and have an even width. The weld beads must be the right size and shape, and in the right place according to the AWS welding symbols. No undercut or overlap should be present. The toes of each fillet should show good fusion with the base metal. The ripples should be even, bullet-shaped, and have no low spots.

## Assigned Job 13-9
# Welding a V-Groove Butt Joint and an Outside Corner Joint in the Overhead Welding Position

### Objectives:
In this job, you will learn to make an acceptable V-groove weld on a butt joint and an outside corner joint in the overhead welding position.

**Note**

Do not attempt this job until you have read all safety precautions, satisfactorily completed the *Shielded Metal Arc Welding (SMAW) Safety Test*, and been approved by your instructor.

1. You will need five pieces of low-carbon steel that measure $1/4'' \times 1\,1/2'' \times 6''$ (6.4 mm × 40 mm × 150 mm). You will also need six $1/8''$ (3.2 mm) diameter E6012 electrodes.

2. Answer the following questions by referring to Figures 11-6 and 11-9 in the text:
   A. What amperage range is suggested for the E6012 electrode? _____–_____ amperes.
   B. What DC polarity is used with an E6013 electrode? _____

3. Set the welding machine amperage at the lower end of the suggested amperage range. Set the correct DC polarity.

4. Flame cut or grind one edge on four pieces to create a 35° bevel.

**Caution**

Wear approved grinding goggles whenever grinding.

5. Place two pieces together to form a V-groove butt joint. Tack weld the joint in three places. The tack welding may be done in the flat welding position.

6. Place the weldment in a welding positioner. Make a practice weld in the overhead welding position to set the welding machine for the best results. The correct electrode angles to use while welding a V-groove butt joint in the overhead welding position are the same as those used in the flat or vertical welding positions.

7. Clean the weld beads with a chipping hammer and a wire brush. Read the weld beads and make whatever changes are required.

**Caution**

Always wear approved chipping goggles when chipping or wire brushing the slag from a weld bead.

8. Arrange the remaining three pieces as shown in the following drawing.

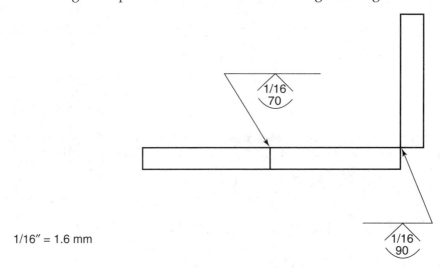

1/16″ = 1.6 mm

9. Tack weld each joint in three places. The tack welding should be done in the weld groove and may be performed in the flat welding position.

10. Place the weldment into a welding positioner. The weldment is turned as required to make each weld in the overhead welding position.

11. Make the V-groove butt welds as shown by the drawing and AWS symbols.

12. Clean each weld bead with a chipping hammer and wire brush.

## Inspection:

All weld beads should be straight and have an even width. There should be full penetration. The weld beads must be the right size and shape, and in the right place according to the AWS welding symbols. No undercut or overlap should be present. The toes of each fillet should show good fusion with the base metal. The ripples should be even, bullet-shaped, and have no low spots.

Name: _____  Date: _____

Class: _____  Instructor: _____

Lesson Grade: _____  Instructor's Initials: _____

## Lesson 14
# Surfacing

## Objectives:

You will be able to discuss the purposes of surfacing various industrial parts and materials. You also will be able to describe how surfacing electrodes are selected and identified, and how surfacing materials are applied to a base metal using the SMAW process.

## Instructions:

Read Chapter 14 and study Figures 14-1 through 14-12. Then, answer or complete the following questions.

_____ 1. If a surface wears by constant rubbing, the wear is called ____.
   A. corrosion
   B. fatigue
   C. abrasion
   D. metal transfer
   E. thermal corrosion

2. What polarity is generally recommended for use with surfacing electrodes when using SMAW equipment?

   _____

3. The surfacing method that sprays high-temperature material onto a surface is called ____ ____.

   _____

4. What cleaning method is generally used on aluminum?

   _____

5. *True or False?* Deep penetration of the surfacing material is highly recommended.

_____ 6. The process of adding a more-easily welded metal to a harder-to-weld metal is called ____.
   A. buttering
   B. buildup
   C. cladding
   D. hardfacing
   E. All of the above.

7. It is cheaper to surface items such as power shovel buckets or stone crushing hammers than it is to replace them. Name two other items that would be cheaper to repair by resurfacing than to replace.

   _____

   _____

8. *True or False?* Since a long arc is recommended for surfacing, a higher-than-normal amperage is used.

_____ 9. How do Rockwell hardness and Brinell hardness testing differ?
A. Rockwell hardness testing is done on softer metals.
B. Brinell hardness testing is done on softer metals.
C. Brinell hardness testing is done on harder metals.
D. Rockwell tests are done most often on copper, aluminum, and bronze.
E. Brinell tests are done on high-carbon steel.

10. ____ and ____ are the chemical elements in a CoCr-A surfacing electrode.

_____

_____ 11. The second part of a surfacing electrode specification indicates ____.
A. whether the rod is to be used as an electrode or a welding rod
B. the polarity for the electrode
C. the welding positions approved for the electrode
D. the filler metal alloy
E. None of the above.

12. *True or False?* A Brinell hardness of 380 is harder than a Rockwell 60 hardness.

_____ 13. What classification of surfacing electrode is recommended for use on bearing and gear surfaces?
A. Fe5.
B. FeMn.
C. CoCr.
D. CuAl.
E. NiCr.

_____ 14. The characteristic of a metal that allows it to withstand hammering blows is ____.
A. hot hardness
B. corrosion resistance
C. metal-to-metal wear resistance
D. abrasion resistance
E. impact strength

15. Parts must be thoroughly cleaned before surfacing. Name three methods used for cleaning before surfacing.

_____

_____

_____

16. On some metals, ____ may be required to ensure a better bonding between the surfacing material and the base metal.

_____

17. *True or False?* Surfacing electrodes are classified by the main alloying ingredients in them.

18. A surfacing electrode is held at approximately a(n) ____ work angle and a(n) ____ drag travel angle.

_____

_____ 19. What technique is used to create a wide weld bead using SMAW?
A. Shorter arc length.
B. Circular motion
C. Forward and back oscillating motion
D. Increase the travel angle

20. *True or False?* Surfacing is only done when it is necessary to harden a surface.

## Assigned Job 14-1
# Hardfacing Mild Steel with Hardfacing Electrodes and SMAW Equipment

## Objective:
In this job, you will learn to apply hardfacing material to the surface of a low-carbon steel plate.

> **Note**
> Do not attempt this job until you have read all safety precautions, satisfactorily completed the *Shielded Metal Arc Welding (SMAW) Safety Test*, and been approved by your instructor.

1. You will need two pieces of low-carbon steel that measure 1/4″ × 3″ × 6″ (6.4 mm × 75 mm × 150 mm). Assume that your metal is the worn cutting edge of a plow that is used on a farm. Refer to Figure 14-9 to determine what type of hardfacing electrode you would choose to resurface the worn plow edge.

2. Answer the following questions:
   A. What type of electrode is recommended for this job? _____
   B. What is the hardness of the recommended electrode? _____
   C. If you use a 1/8″ (3.2 mm) diameter electrode, what is the recommended amperage range? Use the amperage for a 1/8″ (3.2 mm) diameter E6012 electrode, as shown in Figure 11-9 in the text. _____–_____ amperes.
   D. What polarity is suggested for surfacing electrodes used with SMAW equipment? _____
   E. What is the suggested work angle? _____
   F. What is the suggested travel angle? _____

3. Set the welding machine amperage for the middle of the suggested amperage range. Set the correct DC polarity.

4. With soapstone or chalk, draw lines on both surfaces of each piece of metal, as shown in the following drawing.

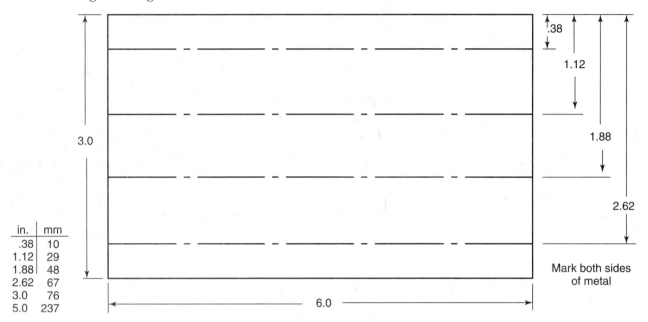

| in. | mm |
|-----|-----|
| .38 | 10 |
| 1.12 | 29 |
| 1.88 | 48 |
| 2.62 | 67 |
| 3.0 | 76 |
| 5.0 | 237 |

5. Use one piece for practice. Using a long arc, lay three stringer beads on one side of the metal.

6. After each weld bead is laid, clean it with a chipping hammer and a wire brush.

**Caution**

Always wear approved chipping goggles when chipping or wire brushing the slag from a weld bead.

7. Read each weld bead as it is laid. Make any necessary corrections to the machine amperage and the way you use the electrode.

8. On the reverse side of your practice piece, lay three weld beads that are about 3/4" (19 mm) wide. Two may be done using a weaving motion. One wide weld bead must be made using the circular motion shown in Figure 14-11 and described in the Surfacing with Shielded Metal section of the text.

9. On the second piece, repeat steps 5 through 8 for a grade.

## Inspection:

The surfacing beads should be similar in appearance to regular shielded metal arc welding beads. The weld beads should be straight and even in width. The ripples should be evenly spaced. The weld bead should be convex in shape, with no low spots.

# Section 3
# Gas Metal and Flux Cored Arc Welding

## Gas Metal and Flux Cored Arc Welding Safety Test

## Objective:

You will be able to discuss the potential safety hazards of GMAW and FCAW, and describe the safety precautions required when working with this equipment.

## Instructions:

Always follow safe practices when welding. If you have safety questions or concerns, ask your instructor. This test does not include questions about every safety topic, but is intended to highlight key items. Review Chapters 15 and 16. Pay special attention to the section on protective clothing and equipment in Chapter 15. Then, answer or complete the following questions.

_____   1.   Which metal transfer type produces the most spatter?
           A.   Short circuiting.
           B.   Globular.
           C.   Spray.
           D.   Pulsed spray.

2.   What input voltage is required to operate most GMAW machines?

    _____

3.   *True or False?* When welding in a closed area, excellent ventilation must be used.

_____   4.   All of the following shielding gases are heavier than air and will displace oxygen, *except* ____.
           A.   argon
           B.   carbon dioxide
           C.   helium
           D.   a mixture of 75% argon and 25% carbon dioxide

5.   The cylinder ____ must be protected when moving a gas cylinder.

    _____

6.   *True or False?* Pressures in a gas cylinder can be more than 7500 psi (52 MPa).

_____    7.  Which of the following filter lenses may be used for GMAW?
              A.  #2.
              B.  #8.
              C.  #12.
              D.  #18.

8.  Following is a list of clothing and other items. Next to each item, mark *Y* if the item is recommended when GMAW or FCAW. Mark *N* if the item is not recommended when welding.

_____    Fire-resistant clothing.

_____    T-shirt.

_____    Cuffs on pants.

_____    Leathers.

_____    Cap.

_____    Welding helmet.

_____    Gloves.

_____    Open-collar shirt.

9.  The steps required to shut down a GMAW or FCAW station are listed below, but are not in the correct order. Put the steps into the correct order by writing the numbers 1–7 in the blanks.

_____    A.   Turn off the welding machine.

_____    B.   Readjust the flowmeter to allow shielding gas to flow.

_____    C.   Coil each cable separately.

_____    D.   Turn off the wire feeder.

_____    E.   Close the cylinder valve.

_____    F.   Press the purge switch to remove all shielding gas from the hose.

_____    G.   Adjust the flowmeter so that shielding gas does not flow.

10.  *True or False?* Adequate ventilation or a fume exhaust is required for FCAW self-shielded process, but is not necessary for the GMAW or FCAW gas-shielded processes.

# Lesson 15
# GMAW and FCAW: Equipment and Supplies

## Objective:

You will be able to describe the equipment used in GMAW and discuss the different types of metal transfer.

## Instructions:

Read Chapter 15 and study Figures 15-1 through 15-28. Then, answer or complete the following questions.

1. What is another name for direct current reverse polarity?

   _____

_____ 2. What type of shielding gas is usually used for globular transfer?
   A.   Argon.
   B.   Argon with 2% oxygen.
   C.   Carbon dioxide.
   D.   Helium.

3. A number ____ shade lens is recommended for GMAW.

   _____

4. What happens when the hole in the contact tube becomes enlarged due to wear? What is done to correct the problem?

   _____

   _____

   _____

5. What term is used by the American Welding Society to describe the process with a continuously fed electrode that uses a shielding gas and no flux?

   _____

_____ 6. In GMAW, when the welding arc becomes longer, the current ____.
   A.   decreases
   B.   increases
   C.   stays the same

7. Label the parts of the welding outfit shown in the following figure:

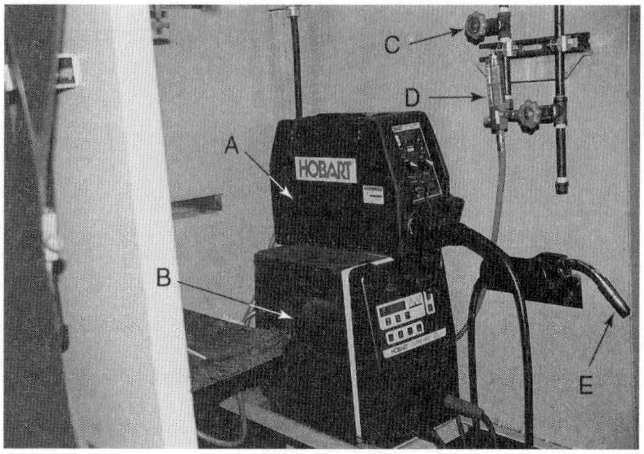

A. _____

B. _____

C. _____

D. _____

E. _____

8. When a flowmeter with a ball enclosed in a clear tube is used, at what point on the ball is the flow of shielding gas indicated or judged?

_____

9. *True or False?* The electrodes used in GMAW or FCAW are called nonconsumable electrodes because they are not used up.

_____     10.  Which types of metal transfer do not put much heat into the work and can be used in all positions?
A     Short circuiting transfer and globular transfer.
B.     Globular transfer and pulsed spray transfer.
C.     Spray transfer and short circuiting transfer.
D.     Pulsed spray transfer and short circuiting transfer.

11. List four advantages of GMAW as compared to shielded metal arc welding.

_____

_____

_____

_____

_____     12.  What type of shielding gas will give deep penetration and undercut when welding on steel?
A.     Helium.
B.     Argon.
C.     Argon with 2% oxygen.
D.     $CO_2$.

Name _____

13. Label the indicated parts of the wire feeder shown in the following figure:

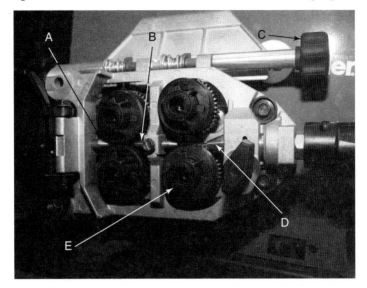

A. _____     D. _____

B. _____     E. _____

C. _____

14. *True or False?* Flux cored arc welding always requires a shielding gas.

_____    15.  Increasing the inductance on a GMAW machine ____.
                 A.  reduces the amount of heat that enters the work
                 B.  increases the transition current
                 C.  reduces spatter during short circuiting transfer
                 D.  increases wire feed speed

16. Pulsed spray transfer has two different currents. Name them.

_____

_____

_____    17.  How many rolls are driven on a wire feeder with two rolls?
                 A.  One.
                 B.  Two.
                 C.  Three.
                 D.  Four.

_____    18.  The following drawing illustrates which type of metal transfer?
               A.  Short circuiting transfer.
               B.  Globular transfer.
               C.  Spray transfer.
               D.  None of the above.

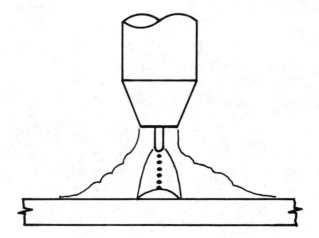

_____    19.  Which of the following gases is not used as a shielding gas?
               A.  Argon.
               B.  Hydrogen.
               C.  Carbon dioxide.
               D.  Helium.

20.  *True or False?* Most GMAW machines have a 60% or 100% duty cycle and require 240-volt power.

## Lesson 16
# GMAW and FCAW:
# Equipment Assembly and Adjustment

## Objectives:
You will be able to describe how GMAW and FCAW equipment is assembled and adjusted. You will also be able to demonstrate how to select an electrode and the correct shielding gas for GMAW and FCAW.

## Instructions:
Read Chapter 16 and study Figures 16-1 through 16-23. Then, answer or complete the following questions.

1. A ball-float flowmeter must be installed in a(n) _____ position.

_____

_____ 2. If the drive rolls in the wire feeder put too much pressure on a flux cored wire, which of the following will occur?
   A. The drive rolls will slip over the wire.
   B. The electrode wire will be crushed or deformed.
   C. The electrode will feed at a faster rate.
   D. The contact tube will not make good electrical contact with the electrode wire.

_____ 3. What is the name of the pliers used to cut electrode wire, and remove and install contact tubes and gas nozzles?
   A. Channel locks.
   B. Needle nose pliers.
   C. Multitool.
   D. Welpers.

4. What happens when the purge switch is pressed?

_____

_____

5. Describe the following electrodes.

   **E80T–1**

   Electrode type: _____

   Tensile strength: _____

   Is $CO_2$ shielding required? _____

**ER110S–1**

Electrode type: _____

Tensile strength: _____

What does the "–1" indicate? _____

6. Which flux-cored electrodes require shielding gas? List the dash numbers.

_____

7. The hose carrying shielding gas is connected to one of two pieces of equipment, depending on the equipment you are using. Name them.

_____

_____

*For Questions 8–12, match the base metals listed at left with the recommended shielding gas for spray transfer from the list at the right. An answer may be used more than once.*

_____ 8. Zirconium.

_____ 9. Low-alloy steel.

_____ 10. Aluminum.

_____ 11. Nickel.

_____ 12. Stainless steel.

A. Helium.

B. Argon.

C. $CO_2$.

D. Argon with 2% oxygen.

E. 50% argon, 50% $CO_2$.

F. 90% helium, 7.5% argon, 2.5% $CO_2$.

_____ 13. Which of the following conditions is not a cause for a bird's nest?
    A. Tension of the drive rolls is too low.
    B. Voltage set is too low.
    C. Welding gun is held too close to work when starting the welding arc.
    D. Wire feed speed is too fast.

*For Questions 14–19, write down at least one electrode that is recommended for gas metal arc welding each of the metals listed.*

14. Nickel: _____

15. Low-alloy steel: _____

16. Aluminum: _____

17. Copper alloys: _____

18. Carbon steel: _____

19. Stainless steel: _____

## Assigned Job 16-1
# Learning to Use the Wire Feeder and Welding Gun

## Objectives:

In this job, you will learn the parts of a wire feeder and a welding gun and how to load electrode wire into the wire feeder and gun. You will also learn about the different switches and adjustments on the wire feeder and gun, and how to calculate the feed rate of the electrode wire.

**Note**
Do not attempt this job until you have read all safety precautions, satisfactorily completed the *Gas Metal and Flux Cored Arc Welding Safety Test*, and been approved by your instructor.

1.  Your instructor will specify a piece of equipment to use for this job. Examine the equipment and answer the following questions.
    A.  Who is the manufacturer of the wire feeder? _____
    B.  How many drive rolls are there? _____
    C.  Is there a wire feed speed control? _____
    D.  Is there an inch switch? _____
    E.  Is there a purge switch? _____

2.  Remove the nozzle and contact tube from the gun. Refer to the Welding Gun section in Chapter 16.
    A.  Is the nozzle clean on the inside? _____ If the nozzle is not clean, use a wire brush to clean it.
    B.  Is the contact tube worn? _____ If it is, replace the contact tube.
    C.  Reinstall the contact tube and nozzle onto the gun. Remember to check these regularly while welding.

3.  Swing the top roll(s) out of the way. Turn the spool of wire to pull the wire out of the cable. Follow the directions in the Wire Feeder section of Chapter 16 to feed the electrode back into the cable leading to the gun. Do not turn on the machine.

4.  How is removing a bird's nest different from the operation in step 3?

    _____

    _____

    _____

5.  Look at the knob on the wire feeder that adjusts the tension on the upper roll(s). Notice how it is used. Adjust the tension as discussed in the Wire Feeder section of Chapter 16 to apply the right tension to the electrode wire.

6. You should now be familiar with the wire feeder and gun. Ask your instructor to look over the equipment and verify that everything is in working condition.
   Instructor's initials: _____

7. Turn the power on. Without striking an arc, pull the trigger on the weld gun. Does the electrode feed smoothly, with no hesitation? _____
   If the answer is no, make proper adjustments until the electrode does feed smoothly.

8. You will need a watch with a second hand for this step. Adjust the wire feed speed to the halfway position. Cut the electrode off flush with the end of the nozzle. Pull the trigger on the gun and hold it down for exactly ten seconds. Release the trigger. Measure the length of electrode wire from the end of the nozzle to the end of the electrode wire.
   What is the length? _____

9. Multiply the length of wire you measured by six (10 seconds × 6 = 60 seconds). What is the answer? _____ This is the wire feed rate, the number of inches (meters) of electrode wire that feed per minute while welding at this setting.

## Inspection:

   When you pull the trigger or press the inch switch on the weld gun, the electrode wire should be pulled off the spool, fed through the wire feeder, and into the cable leading to the gun. The electrode should pass smoothly at a constant speed through the cable to the gun and contact tube.

Name: _____    Date: _____

Class: _____    Instructor: _____

Lesson Grade: _____    Instructor's Initials: _____

## Assigned Job 16-2
# Connecting the Shielding Gas Supply and Adjusting Gas Flow

## Objectives:

In this job, you will learn to connect shielding gas to the welding power supply or wire feeder. You will also learn to properly adjust the shielding gas flow. Review the Regulators and Flowmeters section in Chapter 15 and the Adjusting the Shielding Gas Flowmeter section in Chapter 16.

> **Note**
> Do not attempt this job until you have read all safety precautions, satisfactorily completed the *Gas Metal and Flux Cored Arc Welding Safety Test*, and been approved by your instructor.

1. Your instructor will specify a piece of equipment to use. Examine the equipment and answer the following questions.

2. Who is the manufacturer of the regulator on your equipment? If you cannot find the manufacturer's name on the regulator, or if there is no regulator on your equipment, skip this question.

   _____

3. Who is the manufacturer of the wire feeder? If your wire feeder is internal to the welding machine, name the manufacturer of your welding machine.

   _____

4. Does the equipment you are using have a cylinder of shielding gas at the welding station? ____

5. What type of gas is currently connected to your equipment?

   _____

6. Is the flowmeter a ball-float type or a dial-indicator type (similar to the pressure gauge used on oxyfuel gas equipment)? _____

7. Is there a purge switch on the equipment (either the wire feeder or power supply)? ____ On which piece of equipment is it?

   _____

8. Examine the hose that connects the flowmeter to the welding machine or wire feeder. Is there any evidence of wear? ____ If there is, inform your instructor.

9. Turn the knob or adjusting screw on the flowmeter so that no gas will flow. Disconnect the gas hose from the wire feeder or welding power supply. Examine the fitting. Notice what the fitting on the welding machine looks like. Gas fittings are the same on almost all welding equipment. Reattach the hose to the wire feeder or power supply.

10. With the flowmeter still adjusted so that no shielding gas will flow, open the cylinder valve (if you have a manifold, open the manifold valve).

11. Follow the instructions in the Adjusting the Shielding Gas Flowmeter section of Chapter 16 to set a gas flow of 10 cfh (5 Lpm).

12. Change the flow rate to 20 cfh (9 Lpm).

## Inspection:

You should now be able to connect a supply of shielding gas from a cylinder or manifold to a GMAW or FCAW welding outfit and be able to adjust the flow of shielding gas.

# Lesson 17
# GMAW and FCAW:
# Flat Welding Position

## Objectives:

You will be able to demonstrate how to weld using gas metal arc and flux cored arc welding processes in the flat welding position. You will be able to describe proper gun angles and electrode extension, and how to prevent defects.

## Instructions:

Read Chapter 17 and study Figures 17-1 through 17-18. Complete the following questions.

_____    1. Which of the following will cause the ripples in the weld bead to be unevenly spaced?
  A. A gas flow that is too high.
  B. A wire feed that is too high.
  C. A voltage that is too low.
  D. Moving the gun at uneven speeds.

2. *True or False?* Watching the size of the weld pool will determine the correct travel speed.

_____    3. Why is backing used when making a weld?
  A. To hold two tack welded pieces together.
  B. To obtain consistent penetration.
  C. To prevent undercut on a fillet weld.
  D. To maintain a proper width on a butt weld.

_____    4. Because the contact tube is not visible, which of these distances do you watch and control while welding?
  A. Nozzle-to-work distance.
  B. Electrode extension distance.
  C. Contact tube–to-work distance.

5. When making a fillet weld, what is the proper work angle?

  _____

6. When making a fillet weld with the backhand method, what is the proper drag angle?

  _____

7. Name two things that can be done to reduce spatter.

  _____

  _____

_____    8.  Which of the following produces the best penetration?
         A.  Forehand welding.
         B.  Backhand welding.
         C.  Welding with the gun straight up and down.

_____    9.  Which of the following produces less spatter?
         A.  Forehand welding.
         B.  Backhand welding.
         C.  Welding with the gun straight up and down.

10.  Name the distances shown in the following figure:

    A.  _____

    B.  _____

    C.  _____

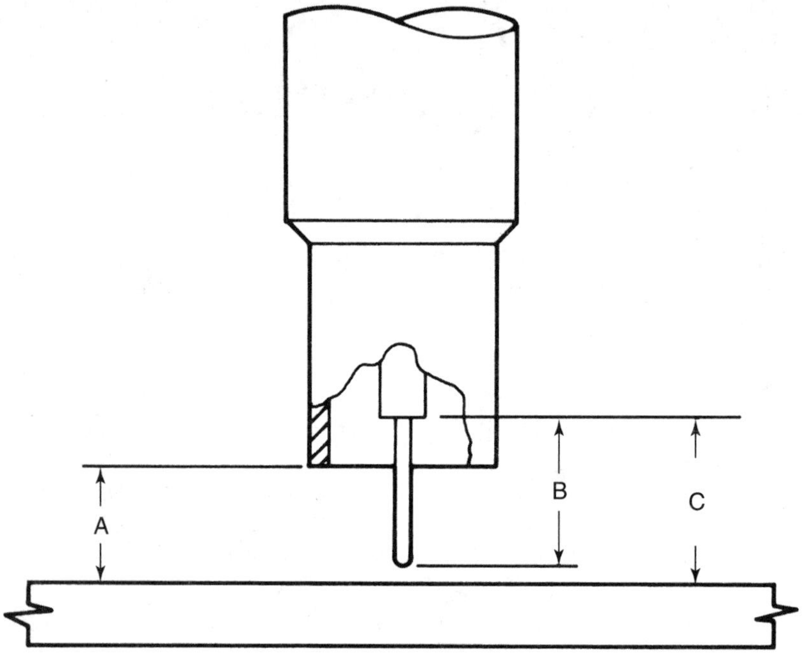

11.  How long should the electrode extension be when using GMAW and short circuiting transfer?

_____

12.  How long should the electrode extension be when using self-shielding FCAW?

_____

_____    13.  What determines the welding arc length in GMAW?
         A.  The voltage set on the welding machine.
         B.  The wire feed speed set on the wire feeder.
         C.  The distance the welder holds the gun from the work.

14.  Briefly explain how to fill the weld pool.

_____

_____

15.  _True or False?_ The welding arc should remain right in the center of the weld pool.

16.  Describe what a weld bead will look like if the travel speed is too fast.

_____

_____

Name: _____   Date: _____

Class: _____   Instructor: _____

Lesson Grade: _____   Instructor's Initials: _____

## Assigned Job 17-1
# Making Stringer and Weave Beads

## Objective:

In this job, you will learn to make stringer and weave beads using short circuiting transfer. This job is to be done using GMAW.

> **Note**
> Do not attempt this job until you have read all safety precautions, satisfactorily completed the *Gas Metal and Flux Cored Arc Welding Safety Test*, and been approved by your instructor.

1. Obtain several pieces of low-carbon steel that measure 1/8″ × 4″ × 6″ (3.2 mm × 100 mm × 150 mm).

2. Mark four straight lines on one side of each piece of metal with soapstone or chalk. The lines should be about 1″ (25 mm) apart and 1/2″ (13 mm) from the edge. The lines should run the 6″ (150 mm) length of the metal.

3. Before setting up the welding machine, answer the following questions:
   A. What type of electrode is recommended for welding low-carbon steel using GMAW?
   
   _____
   
   B. A .035″ (0.9 mm) diameter electrode wire is recommended, although other diameters can be used. What arc voltage is recommended for .035″ (0.9 mm) diameter electrode when using short circuiting transfer? Refer to Figure 16-18 in the text. _____ volts
   C. What wire feed speed is recommended for a .035″ (0.9 mm) electrode? Refer to Figure 16-18. _____ in/min (_____ M/min)
   D. What choices of shielding gases do you have when welding mild steel? Refer to Figure 16-11.
   
   _____
   
   _____
   
   E. What shielding gas flow rate should be used for welding 1/8″ mild steel in the flat welding position? Refer to Figure 17-2. _____

4. Make a safety check of your welding outfit before continuing this job.

5. Mount the proper electrode onto the wire feeder and feed the wire to the welding gun. Refer to the headings related to the wire feeder and setting up the equipment in Chapters 16 and 17. Also refer to Assigned Job 16-1. Make sure you are using the correct drive wheels and correct contact tube for the diameter electrode you are using.

6. Connect the shielding gas and set the flow rate. Refer to the sections related to the equipment and its proper setup in Chapter 16 of the text. Also see Assigned Job 16-2.

7. Set the desired voltage on the welding machine.

8. Adjust the wire feeder to obtain the correct wire feed speed. Refer to the Wire Feeder section in Chapter 16 and Assigned Job 16-1.

9. Place the metal you will be welding onto your work table or into a ground clamp, so that it is electrically connected to the workpiece lead of the welding machine.

10. Make sure you are wearing the proper clothing for welding. Refer to the heading Protective Clothing and Equipment in Chapter 15. Check the shade of the lens in your helmet. It should be a #12 lens. After completing this step, you will be ready to begin welding. Refer to the Preparing to Weld section in Chapter 17.

11. Hold the welding gun over the place you will begin welding. Grasp the gun so you will be welding using the backhand method. The electrode extension should be 1/4"–1/2" (6 mm–13 mm). There should be a gap of about 1/16" (1.5 mm) between the end of the electrode and the work.

12. Lower your helmet and pull the trigger on the gun.

13. As soon as the weld pool develops, begin moving forward. Keep the welding arc just in front of the center of the pool. GMAW is faster than SMAW and much faster than GTAW or oxyfuel gas welding.

14. Weld two stringer beads in a straight line along the metal. Fill the weld pool before you stop. You may need to adjust the voltage and/or wire feed speed to obtain a high-quality weld. You can also adjust the slope if your machine has a slope control.

15. Continue making stringer beads until you are creating good weld beads. See the inspection information at the end of this job.

16. Next, weld weave beads until you make good-quality weave beads. These weave beads should be about twice as wide as a stringer bead.

17. Once you have practiced welding stringer and weave beads, weld two stringer beads and two weave beads on a single plate. Show these to your instructor for a grade.

## Inspection:

Each weld bead should be straight, with ripples that are evenly spaced. The width of the weld should remain the same from beginning to end. The weld pool at the stopping point should be filled. There should not be a lot of spatter on the plate.

## Assigned Job 17-2
# Making Stringer Beads Using Spray Transfer

## Objective:

In this job, you will learn to make stringer beads using GMAW and spray transfer on mild steel.

> **Note**
> Do not attempt this job until you have read all safety precautions, satisfactorily completed the *Gas Metal and Flux Cored Arc Welding Safety Test*, and been approved by your instructor.

1. Obtain several pieces of low-carbon steel that measure 3/16″ × 4″ × 6″ (4.8 mm × 100 mm × 150 mm).

2. Mark four straight lines on one side of each piece of metal with soapstone or chalk. The lines should be about 1″ (25 mm) apart and 1/2″ (13 mm) from the edge. The lines should run the 6″ (150 mm) length of the metal.

3. Before setting up the welding machine, answer the following questions.
   A. What diameter electrode will you be using? A .035″ (0.9 mm) electrode diameter is recommended. _____
   B. What arc voltage is recommended for the diameter electrode you will be using when using spray transfer? Refer to Figure 16-18 in the text. _____
   C. What wire feed speed is recommended? Refer to Figure 16-18. _____
   D. What type of shielding gas will you be using for welding mild steel? Refer to Figure 16-12. _____
   E. What shielding gas flow rate should be used when welding in the flat welding position? Refer to Figure 17-2. _____

4. Follow steps 4–10 of Assigned Job 17-1 to set up the welding station. After completing this step, you will be ready to begin welding.

5. Hold the welding gun over the place you will begin welding. Grasp the gun so you will be welding using the backhand method. The electrode extension should be 1/2″–1″ (13 mm–25 mm). There should be about 1/16″ (1.5 mm) gap between the end of the electrode and the work.

6. Lower your helmet and pull the trigger on the gun.

7. In spray transfer, the weld pool develops very quickly. Almost as soon as you pull the trigger, you will need to begin moving forward. Keep the welding arc just in front of the center of the weld pool. Spray transfer requires you to move at about twice the speed you did when using short circuiting transfer. Do not be afraid of moving too fast. Fill the weld pool when you reach the end of each weld.

8. Adjust the voltage and wire feed so you can make good-quality welds.

9. Continue making stringer beads until you are making good welds. When you are satisfied with the quality of your welds, make four stringer beads on one piece of metal and show them to your instructor.

## Inspection:

Spray transfer usually makes a weld bead with an excellent appearance. Each weld bead should be straight with evenly spaced ripples. The width of the weld bead should remain the same from beginning to end, and the weld pool at the stopping point should be filled. There should be no spatter.

Name: _____ Date: _____

Class: _____ Instructor: _____

Lesson Grade: _____ Instructor's Initials: _____

## Assigned Job 17-3
# Welding a Lap Joint in the Flat Welding Position

## Objective:

In this job, you will learn to weld a fillet weld on the lap joint in the flat welding position. You can perform this job with GMAW and FCAW.

> **Note**
> Do not attempt this job until you have read all safety precautions, satisfactorily completed the *Gas Metal and Flux Cored Arc Welding Safety Test*, and been approved by your instructor.

1. Obtain four pieces of low-carbon steel that measure 3/16″ × 1 1/2″ × 6″ (4.8 mm × 40 mm × 150 mm).

2. Answer the following questions before setting up the welding machine.
   A. Which welding process will you be using, GMAW or FCAW? _____
   B. If using GMAW, what type of metal transfer will you be using? _____
   C. What electrode diameter will you be using? _____
   D. What is the recommended arc voltage and wire feed speed for the type of metal transfer and electrode diameter you are using?
      Arc voltage: _____
      Wire feed speed: _____
   E. What is the recommended type of shielding gas and shielding gas flow rate?
      Shielding gas: _____
      Flow rate: _____
   F. What is the recommended electrode extension? _____
   G. What is the proper work angle? Refer to Figure 17-12 in the text. _____
   H. What is the proper travel angle? Refer to Figure 17-12. _____

3. Set up the equipment and verify that it is operating properly.

4. Place one piece of metal on the other so that it overlaps about 3/4″ (19 mm). Tack weld in three places on each side of the joint.

5. Support the tack-welded parts so the joint is in the flat welding position.

6. Weld a fillet weld on both sides of the metal, as shown in the following drawing. Both welds are to be done in the flat welding position. Refer to Figure 17-12 for the proper gun angles.

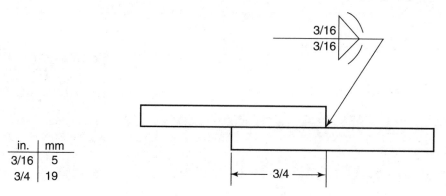

| in. | mm |
|-----|-----|
| 3/16 | 5 |
| 3/4 | 19 |

7. Tack weld the other two pieces of metal and weld two fillet welds as shown in the previous figure. Weld the joints in the flat welding position.

## Inspection:

Each fillet weld should be straight, slightly convex, and have evenly spaced ripples. The fillet welds should have equal legs of 3/16" (5 mm). There should be no undercut or overlap.

## Assigned Job 17-4
# Welding a T-Joint in the Flat Welding Position

## Objective:

In this job, you will learn to weld a fillet weld on a T-joint in the flat welding position. You can perform this job with GMAW and FCAW.

> **Note**
> Do not attempt this job until you have read all safety precautions, satisfactorily completed the *Gas Metal and Flux Cored Arc Welding Safety Test*, and been approved by your instructor.

1. Obtain four pieces of low-carbon steel that measure 3/16″ × 1 1/2″ × 6″ (4.8 mm × 40 mm × 150 mm), and two pieces that measure 3/16″ × 3″ × 6″ (4.8 mm × 75 mm × 150 mm).

2. Answer the following questions before setting up the welding machine.
   A. Which welding process will you be using, GMAW or FCAW? _____
   B. If using GMAW, what type of metal transfer will you be using? _____
   C. What electrode diameter will you be using? _____
   D. What is the recommended arc voltage and wire feed speed for the type of metal transfer and electrode diameter you are using? Refer to Figure 16-18 in the text.
      Arc voltage: _____
      Wire feed speed: _____
   E. What is the recommended type of shielding gas and flow rate? Refer to Figures 16-11, 16-12, and 17-2.
      Shielding gas: _____
      Flow rate: _____
   F. What is the recommended electrode extension? _____
   G. What is the proper work angle? _____
   H. What is the proper travel angle? _____

3. Set up the equipment and verify that everything is operating properly.

4. Place one of the smaller pieces onto a larger piece as shown in the following drawing. Tack weld the joint in three places on each side.

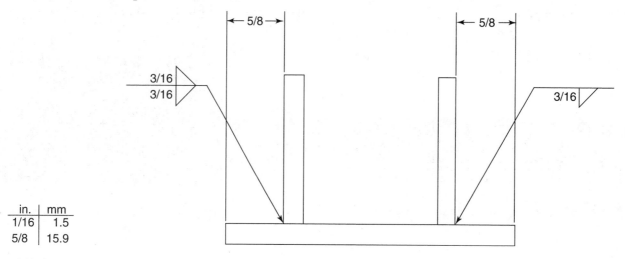

| in. | mm |
|-----|------|
| 1/16 | 1.5 |
| 5/8 | 15.9 |

5. Support the tack-welded parts so the joint is in the flat welding position.

6. Weld a fillet weld on both sides of the metal, as shown in the drawing. Both welds are to be done in the flat welding position.

7. Tack weld the second smaller piece onto the assembly. Weld a fillet weld in the flat welding position as shown in the drawing for this weldment. Inspect these welds, as described in the Inspection: section of this job.

8. Using the remaining material, repeat this welding exercise for a grade. Weld all the joints in the flat welding position.

## Inspection:

Each fillet weld should be straight, slightly convex, and have evenly spaced ripples. The fillet welds should have equal legs of about 3/16″ (5 mm). There should be no undercut or overlap.

## Assigned Job 17-5
# Welding a Square-Groove Butt Joint in the Flat Welding Position

## Objective:

In this job, you will learn to weld a square-groove butt joint in the flat welding position. You can perform this job with GMAW and FCAW.

> **Note**
>
> Do not attempt this job until you have read all safety precautions, satisfactorily completed the *Gas Metal and Flux Cored Arc Welding Safety Test*, and been approved by your instructor.

1. Obtain four pieces of low-carbon steel that measure 1/8″ × 1 1/2″ × 6″ (3.2 mm × 40 mm × 150 mm).

2. Answer the following questions before setting up the welding machine.
   A. Which welding process will you be using, GMAW or FCAW? _____
   B. If using GMAW, what type of metal transfer will you be using? _____
   C. What electrode diameter will you be using? _____
   D. What is the recommended arc voltage and wire feed speed for the type of metal transfer and electrode diameter you are using?
   Arc voltage: _____
   Wire feed speed: _____
   E. What is the recommended type of shielding gas and flow rate?
   Shielding gas: _____
   Flow rate: _____
   F. What is the recommended electrode extension? _____
   G. What is the proper work angle? _____
   H. What is the proper travel angle? _____

3. Set up the equipment and verify that everything is operating properly.

4. Align two pieces of metal so there is a 1/16″ (1.5 mm) gap between them. Weld three tack welds along the joint. After tacking, the gap should still measure 1/16″ (1.5 mm) along its entire length.

5. Position the metal so the weld can be made in the flat welding position. Weld the butt joint as shown in the following drawing. Watch the weld pool as you weld. It must melt both pieces being welded. You should be able to judge the amount of penetration as you weld. Adjust your travel speed to completely penetrate the joint. You may also need to adjust the electrode extension, the voltage, and the wire feed speed.

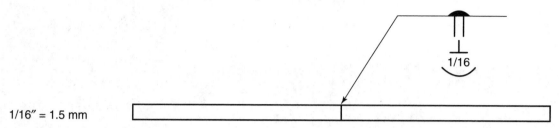

1/16″ = 1.5 mm

6. Tack weld the remaining two pieces and then weld them. After making each weld, examine it and make any corrections necessary. Refer to the Welding a Butt Joint section in Chapter 17.

## Inspection:

Each weld should be straight, slightly convex, and have the ripples evenly spaced. After wire brushing, there should be very little spatter on the plate or on the weld. There should be an even amount of penetration on the back side of the weld.

## Lesson 18
# GMAW and FCAW: Horizontal, Vertical, and Overhead Welding Positions

## Objectives:

You will be able to demonstrate how to weld lap joints, outside corner joints, T-joints, and butt joints in the horizontal, vertical, and overhead welding positions using the gas metal and flux cored arc welding process.

## Instructions:

Read Chapter 18 and study Figures 18-1 through 18-12. Then, answer or complete the following questions.

1. *True or False?* The forehand method provides better penetration than the backhand method.

_____ 2. Which of the following is a correct step to take if the weld pool gets too large when welding out of position?
   A. Increase the wire feed speed.
   B. Change electrode types.
   C. Change shielding gas to helium.
   D. Increase travel speed.

_____ 3. Which of these types of metal transfer can be used when welding out of position?
   A. Short circuiting transfer.
   B. Globular transfer.
   C. Spray transfer.
   D. None of the above.

4. *True or False?* When welding overhead, the shielding gas flow rate must be increased if argon is the shielding gas.

5. *True or False?* The welding gun is usually pointed up when welding vertically up and is also pointed up when welding vertically down.

_____ 6. Which of the following is *not* a reason for welding in the flat welding position?
   A. All types of metal transfer can be used.
   B. The weld pool will not sag or fall from the weld joint.
   C. Helium shielding gas is lighter than other shielding gases.
   D. Welding in the flat position is more comfortable.

7. What is a backing used for?

_____

_____

_____    8. Which welding position is considered to be the most difficult?
A. Flat.
B. Horizontal.
C. Vertical.
D. Overhead.

9. A lap joint often requires that the electrode be pointed more toward the _____ than the _____.

_____

10. *True or False?* A horizontal butt weld in which the molten metal sags down and causes the top part of the weld joint not to be filled is considered acceptable.

Name: _____ Date: _____

Class: _____ Instructor: _____

Lesson Grade: _____ Instructor's Initials: _____

**Assigned Job 18-1**

# Welding a Lap Joint in the Horizontal Welding Position

## Objective:

In this job, you will learn to make a fillet weld on a lap joint in the horizontal welding position. You can perform this job with GMAW and FCAW.

---

**Note**

Do not attempt this job until you have read all safety precautions, satisfactorily completed the *Gas Metal and Flux Cored Arc Welding Safety Test*, and been approved by your instructor.

---

1. Obtain four pieces of low-carbon steel that measure 3/16″ × 1 1/2″ × 6″ (4.8 mm × 40 mm × 150 mm) and two pieces that measure 3/16″ × 3″ × 6″ (4.8 mm × 75 mm × 150 mm).

2. Answer the following questions before setting up the welding machine.
   A. Which welding process will you be using, GMAW or FCAW? _____
   B. If using GMAW, what type of metal transfer will you be using? _____
   C. What electrode diameter will you be using? _____
   D. What is the recommended arc voltage and wire feed speed for the type of metal transfer and electrode diameter you are using? Refer to Figure 16-18 in the text.
      Arc voltage: _____
      Wire feed speed: _____
   E. What is the recommended type of shielding gas and flow rate? Refer to Figures 16-11, 16-12, and 17-2.
      Shielding gas: _____
      Flow rate: _____
   F. What is the recommended electrode extension? _____
   G. What is the proper work angle? _____
   H. What is the proper travel angle? _____

3. Set up the equipment and verify that everything is operating properly.

4. Align the three pieces of metal to be welded, as shown in the following drawing. Tack weld the parts. Tack welding may be done in the flat welding position.

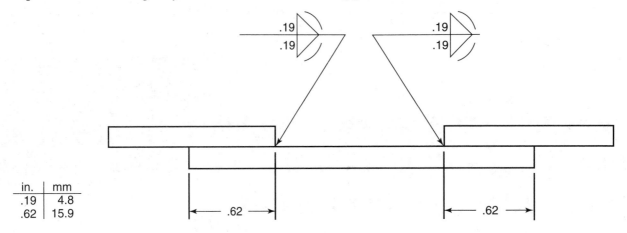

| in. | mm |
| --- | --- |
| .19 | 4.8 |
| .62 | 15.9 |

5. Place the tack-welded parts so the joint to be welded is in the horizontal welding position.

6. Make the fillet welds indicated on the drawing. Reposition the metal as needed, so that all welds are made in the horizontal welding position. Refer to Figure 18-2 in the text for the proper gun angles.

7. Tack weld the other pieces of metal together, and then weld them as shown in the drawing. Weld each joint in the horizontal welding position.

## Inspection:

Each fillet weld should be straight, slightly convex, and have an even width with evenly spaced ripples in the weld bead. There should be no undercut or overlap.

Name: _____     Date: _____

Class: _____     Instructor: _____

Lesson Grade: _____     Instructor's Initials: _____

## Assigned Job 18-2
# Welding a T-Joint in the Horizontal Welding Position

### Objective:

In this job, you will learn to make a fillet weld on a T-joint in the horizontal welding position. You can perform this job with GMAW and FCAW.

> **Note**
> Do not attempt this job until you have read all safety precautions, satisfactorily completed the *Gas Metal and Flux Cored Arc Welding Safety Test*, and been approved by your instructor.

1. Obtain four pieces of low-carbon steel that measure 3/16″ × 1 1/2″ × 6″ (4.8 mm × 40 mm × 150 mm) and two pieces that measure 3/16″ × 3″ × 6″ (4.8 mm × 75 mm × 150 mm).

2. Answer the following questions before setting up the welding machine.
   A. Which welding process will you be using, GMAW or FCAW? _____
   B. What type of metal transfer will you be using? _____
   C. What electrode diameter will you be using? _____
   D. What is the recommended arc voltage and wire feed speed for the type of metal transfer and the electrode diameter you are using? _____
      Arc voltage: _____
      Wire feed speed: _____
   E. What is the recommended shielding gas and flow rate? Refer to Figures 12-11, 12-12, and 13-2 in the text.
      Shielding gas: _____
      Flow rate: _____
   F. What is the recommended electrode extension? _____
   G. What is the proper work angle? _____
   H. What is the proper travel angle? _____
   I. What is the shape of the weld pool when making a fillet weld in the horizontal welding position? _____

3. Set up the equipment and verify that everything is operating properly.

4. Place one of the smaller pieces onto a large piece as shown in the following drawing. Tack weld in three places on each side of the joint.

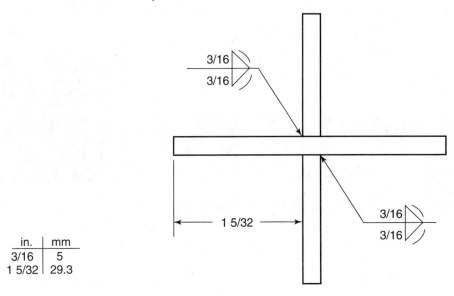

| in. | mm |
|-----|-----|
| 3/16 | 5 |
| 1 5/32 | 29.3 |

5. Place the tack-welded metal so the joint is in the horizontal welding position.

6. Weld a fillet weld on both sides of the metal, as shown on the drawing. Both welds are to be done in the horizontal welding position. Refer to Figure 18-2 in the text for the proper gun angles.

7. Tack weld the second of the smaller pieces onto the assembly. Place the tack-welded metal in a positioner so the joint is in the horizontal position. Weld a fillet weld on each side of the T-joints. Weld each joint in the horizontal welding position. Inspect the welds.

8. Repeat this welding job with the remaining material. Weld all the joints in the horizontal welding position.

## Inspection:

Each fillet weld should be straight, slightly convex, and have evenly spaced ripples. The fillet welds should have equal legs of 3/16" (5 mm). There should be no undercut or overlap.

Name: _____     Date: _____

Class: _____     Instructor: _____

Lesson Grade: _____     Instructor's Initials: _____

## Assigned Job 18-3
# Welding a Square-Groove Butt Joint in the Horizontal Welding Position

## Objective:

In this job, you will learn to weld a square-groove butt joint in the horizontal welding position. You should perform this job first using the GMAW process and short circuiting transfer. You can then repeat the job using FCAW.

> **Note**
>
> Do not attempt this job until you have read all safety precautions, satisfactorily completed the *Gas Metal and Flux Cored Arc Welding Safety Test*, and been approved by your instructor.

1. Obtain six pieces of mild steel that measure 1/8″ × 1 1/2″ × 6″ (3.2 mm × 40 mm × 150 mm). You may also use a backing.

2. Answer the following questions before setting up the welding machine.
   A. What electrode diameter will you be using? _____
   B. What is the recommended arc voltage and wire feed speed for short circuiting transfer and the electrode diameter you are using?
   Arc voltage: _____
   Wire feed speed: _____
   C. What is the recommended shielding gas and flow rate? Refer to Figures 16-11 and 17-2 in the text.
   Shielding gas: _____
   Flow rate: _____
   D. What is the recommended electrode extension? _____
   E. What is the proper work angle? _____
   F. What is the proper travel angle? _____

3. Set up the equipment and verify that everything is operating properly.

4. Align two pieces of metal so that there is a 1/16″ (1.5 mm) gap between them. Make three tack welds along the joint. After tacking, the gap should still measure 1/16″ (1.5 mm) along the entire length. Tack weld a third piece onto the first two, as shown in the following drawing.

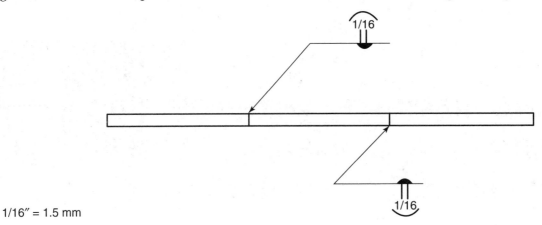

1/16″ = 1.5 mm

5. Place the metal in a positioner so the weld can be made in the horizontal welding position. Weld the butt joint as shown in the drawing. Watch the weld pool as you weld. Do not allow it to sag. If it does begin to sag, change the work angle, increase your travel speed, or decrease the wire feed speed.

   To ensure 100% penetration, the gap at the root of the weld pool should look like a keyhole. Keep this keyhole the same size as you make the weld. You may need to adjust the voltage and wire feed speed to create the keyhole. The keyhole will be smaller than it is in oxyfuel gas or shielded metal arc welding. The edges of both pieces being welded must be molten.

6. Tack weld the three remaining pieces, and then weld them as directed. After making each weld, examine it and make any corrections necessary. Refer to the Welding a Butt Joint section in Chapter 18.

## Inspection:

Each butt weld should be straight, slightly convex, and have evenly spaced ripples. There should not be a lot of sag of the weld onto the lower piece. After wire brushing, there should be very little spatter on the plate or on the weld. The penetration on the back side of the weld should be flush to slightly convex.

## Assigned Job 18-4
# Welding a Lap Joint in the Vertical Welding Position

## Objective:

In this job, you will learn to make a fillet weld on a lap joint in the vertical welding position. You can perform this job with GMAW and FCAW.

> **Note**
> Do not attempt this job until you have read all safety precautions, satisfactorily completed the *Gas Metal and Flux Cored Arc Welding Safety Test*, and been approved by your instructor.

1. Obtain six pieces of low-carbon steel that measure 3/16″ × 1 1/2″ × 6″ (4.8 mm × 40 mm × 150 mm).

2. Answer the following questions before setting up the welding machine.
   A. Which welding process will you be using, GMAW or FCAW? _____
   B. What type of metal transfer is recommended for vertical GMAW? _____
   C. What electrode diameter will you be using? _____
   D. What is the recommended arc voltage and wire feed speed for the type of metal transfer and the electrode diameter you are using?
   Arc voltage: _____
   Wire feed speed: _____
   E. What is the recommended shielding gas and flow rate?
   Shielding gas: _____
   Flow rate: _____
   F. What is the recommended electrode extension? _____
   G. What is the proper work angle? _____
   H. Does the weld pool sag less when welding uphill or downhill? _____
   I. Where is the electrode pointed? _____
   J. What is the shape of the weld pool when making a fillet weld in the vertical welding position? _____

3. Set up the equipment and verify that everything is operating properly.

4.  Align the metal to be welded as shown in the following drawing and tack the weldment together.

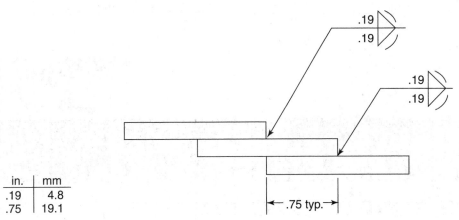

| in. | mm |
|-----|------|
| .19 | 4.8 |
| .75 | 19.1 |

5.  Place the tack-welded parts in a positioner so the joint to be welded is in the vertical welding position.

6.  Weld the fillet welds as indicated on the drawing. Reposition the metal as required so all welds are made in the vertical welding position. Refer to Figure 18-2 in the text for the proper gun angles. Weld in the downhill direction.

7.  Tack weld the other pieces of metal together and then weld them as shown in the drawing. Keep the weld joint in the vertical position.

## Inspection:

Each fillet weld should be straight, slightly convex, have an even width, and evenly spaced ripples in the weld bead. There should be no undercut and very little spatter.

## Assigned Job 18-5
# Welding a T-Joint in the Vertical Welding Position

## Objective:

In this job, you will learn to weld a fillet weld on a T-joint in the vertical welding position. You can perform this job with GMAW and FCAW.

> **Note**
> Do not attempt this job until you have read all safety precautions, satisfactorily completed the *Gas Metal and Flux Cored Arc Welding Safety Test*, and been approved by your instructor.

1. Obtain four pieces of low-carbon steel that measure 1/8″ × 1 1/2″ × 6″ (3.2 mm × 40 mm × 150 mm) and four pieces that measure 3/16″ × 1 1/2″ × 6″ (4.8 mm × 40 mm × 150 mm).

2. Answer the following questions before setting up the welding machine.
   A. Which welding process will you be using, GMAW or FCAW? _____
   B. What electrode diameter will you be using? _____
   C. What is the recommended arc voltage and wire feed speed for the type of metal transfer and electrode diameter you are using?
   Arc voltage: _____
   Wire feed speed: _____
   D. What is the recommended shielding gas and flow rate?
   Shielding gas: _____
   Flow rate: _____
   E. What is the recommended electrode extension? _____
   F. What is the proper work angle? _____
   G. Does the weld pool sag less when you weld uphill or downhill? _____
   H. Where is the electrode pointed? _____
   I. What is the shape of the weld pool when making a fillet weld in the vertical welding position? _____

3. Set up the equipment and verify that everything is operating properly.

4. Place one of the 1/8″ (3.2 mm) pieces on the other 1/8″ (3.2 mm) piece to form the weldment shown in Part A of the following drawing. Tack weld in three places on each side of the joint.

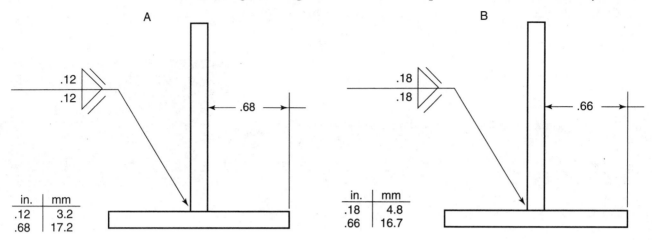

5. Repeat step 4 for the 3/16″ (4.8 mm) pieces.

6. Place the 1/8″ (3.2 mm) tack-welded metal in a positioner, so the joint to be welded is in the vertical welding position.

7. Weld a fillet weld on both sides of the 1/8″ (3.2 mm) metal, as shown in Part A of the drawing. Both welds are to be done in the vertical welding position. Refer to the Welding in the Vertical Welding Position section in Chapter 18 for the proper gun angles.

8. Weld a fillet weld on both sides of the 3/16″ (4.8 mm) metal, as shown in Part B of the drawing. The welds are to be made in the vertical welding position. This weld may require more than one pass to complete. Before making a second weld bead, the first weld bead must be cleaned. Each weld bead must melt into the base metal and the previous weld bead. You can use stringer beads or a weaving motion, but keep the weld pool small so it does not sag.

## Inspection:

Each fillet weld should be straight, slightly convex, and have evenly spaced ripples. The fillet welds should have equal legs. There should be no undercut or overlap. A weld made in the vertical welding position should look as good as a weld done in the flat welding position.

## Assigned Job 18-6
# Welding a Square-Groove Butt Joint in the Vertical Welding Position

## Objective:

In this job, you will learn to weld a square-groove butt joint in the vertical welding position. You can perform this job with GMAW and FCAW.

> **Note**
>
> Do not attempt this job until you have read all safety precautions, satisfactorily completed the *Gas Metal and Flux Cored Arc Welding Safety Test*, and been approved by your instructor.

1. Obtain six pieces of mild steel that measure 1/8″ × 1 1/2″ × 6″ (3.2 mm × 40 mm × 150 mm). You may also use a backing.

2. Answer the following questions before setting up the welding machine.
   A. Which welding process will you be using, GMAW or FCAW? _____
   B. What electrode diameter will you be using? _____
   C. What is the recommended arc voltage and wire feed speed for the type of metal transfer and electrode diameter you are using?
      Arc voltage: _____
      Wire feed speed: _____
   D. What is the recommended shielding gas and flow rate?
      Shielding gas: _____
      Flow rate: _____
   E. What is the recommended electrode extension? _____
   F. What is the proper work angle? _____
   G. What is the proper travel angle? _____

3. Set up the welding equipment and verify that everything is operating properly.

4. Align two pieces of metal so there is a 1/16″ (1.5 mm) gap between them. Weld three tack welds along the joint. After tacking, the gap should still measure 1/16″ (1.5 mm) along the entire length. Tack weld a third piece onto the first two, as shown in the following drawing. Tack welding can be done in the flat welding position.

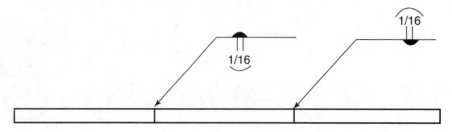

1/16″ = 1.5 mm

5. Place the metal in a positioner so the weld can be made in the vertical welding position. Weld the butt joints as shown in the drawing. The root of the weld pool should look like a keyhole. Keep this keyhole the same size as you make the weld. If you are using a backing, the wire feed speed can be increased slightly. You will not see a keyhole if you are using a backing bar because there is metal behind the joint.

6. Tack weld the remaining pieces and then weld them. After making each weld, examine it and make any necessary corrections.

## Inspection:

Each butt joint weld should be straight, slightly convex, and have evenly spaced ripples. There should be no sagging of the weld bead. Penetration on the back side of the weld should be 100%, and should be consistent along the weld joint. After wire brushing, there should be very little spatter on the plate or on the weld.

## Assigned Job 18-7
# Welding a Lap Joint in the Overhead Welding Position

## Objective:

In this job, you will learn to make a fillet weld on a lap joint in the overhead welding position using short circuiting transfer.

> **Note**
> Do not attempt this job until you have read all safety precautions, satisfactorily completed the *Gas Metal and Flux Cored Arc Welding Safety Test*, and been approved by your instructor.

1. Obtain six pieces of low-carbon steel that measure 3/16″ × 1 1/2″ × 6″ (4.8 mm × 40 mm × 150 mm).

2. Answer the following questions before setting up the welding machine.
   A. Which welding process will you be using, GMAW or FCAW? _____
   B. What type of metal transfer is recommended for overhead GMAW? _____
   C. What electrode diameter will you be using? _____
   D. What is the recommended arc voltage and wire feed speed for the type of metal transfer and electrode diameter you are using?
      Arc voltage: _____
      Wire feed speed: _____
   E. What is the recommended shielding gas and flow rate?
      Shielding gas: _____
      Flow rate: _____
   F. Is the shielding gas flow rate higher or lower when welding in the overhead welding position compared to other positions? _____
   G. What is the recommended electrode extension? _____
   H. What is the proper work angle? _____
   I. Where is the electrode pointed? _____
   J. What is the shape of the weld pool when making a fillet weld in the overhead welding position? _____

3. Set up the equipment and verify that everything is operating properly.

4.  Align two pieces of metal (pieces A and B) as shown in the following drawing. Tack weld each lap joint in three places. This may be done in the flat welding position.

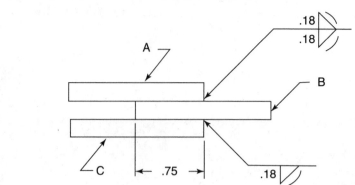

| in. | mm |
|-----|-----|
| .18 | 4.8 |
| .75 | 19 |

5.  Place the tack-welded parts in a positioner so the joint can be welded in the overhead welding position.

6.  Make the fillet welds as indicated on the drawing. Turn the metal as required so each weld is made in the overhead welding position. Keep the weld pool small. There can be no overlap of weld metal onto the surface of piece B, or piece C will not fit tightly.

7.  Tack weld piece C into place. Weld a fillet weld in the overhead welding position.

8.  For a grade, repeat this job with the remaining material.

## Inspection:

Each fillet weld should be straight and slightly convex. The fillet welds should have an even width and evenly spaced ripples in the weld bead. There should be no undercut or overlap.

## Assigned Job 18-8
# Welding an Inside Corner Joint in the Overhead Welding Position

### Objective:

In this job, you will learn to weld a fillet weld on an inside corner joint in the overhead welding position. You can perform this job with GMAW and FCAW.

> **Note**
> Do not attempt this job until you have read all safety precautions, satisfactorily completed the *Gas Metal and Flux Cored Arc Welding Safety Test*, and been approved by your instructor.

1. Obtain six pieces of low-carbon steel that measure 3/16″ × 1 1/2″ × 6″ (4.8 mm × 40 mm × 150 mm) and three pieces that measure 3/16″ × 3″ × 6″ (4.8 mm × 75 mm × 150 mm).

2. Answer the following questions before setting up the welding machine.
   A. Which welding process will you be using, GMAW or FCAW? _____
   B. What electrode diameter will you be using? _____
   C. What is the recommended arc voltage and wire feed speed for the type of metal transfer and electrode diameter you are using?
   Arc voltage: _____
   Wire feed speed: _____
   D. What is the recommended shielding gas and flow rate?
   Shielding gas: _____
   Flow rate: _____
   E. What is the recommended electrode extension? _____
   F. What is the proper work angle? _____
   G. Where is the electrode pointed? _____
   H. What is the shape of the weld pool when making a fillet weld in the overhead welding position? _____

3. Set up the equipment and verify that everything is operating properly.

4. Align and tack weld three pieces of metal to form the weldment shown in the following drawing. Tack welding can be done in the flat welding position.

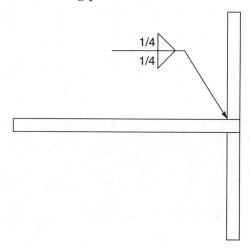

1/4″ = 6.4 mm

5. Place the tack-welded metal in a positioner so the joint is in the overhead welding position.

6. Make the fillet welds as shown on the drawing. Both welds are to be done in the overhead welding position. More than one pass will be needed. Clean each weld bead before making the next one.

7. Repeat this job with the remaining pieces of metal. Evaluate each fillet weld after completing it. Make any changes before making the next weld. Show the final weldment to your instructor for a grade.

## Inspection:

Each fillet weld should be straight, slightly convex, and have evenly spaced ripples. The fillet welds should have equal legs. There should not be any undercut or overlap. A weld made in the overhead welding position should look as good as a weld done in the flat welding position.

## Assigned Job 18-9
# Welding a Square-Groove and V-Groove Butt Joint in the Overhead Welding Position

## Objectives:

In this job, you will learn to weld a square-groove butt joint and a V-groove butt joint on an outside corner joint in the overhead welding position. You can perform this job with GMAW and FCAW.

**Note**

Do not attempt this job until you have read all safety precautions, satisfactorily completed the *Gas Metal and Flux Cored Arc Welding Safety Test*, and been approved by your instructor.

1. Obtain six pieces of mild steel that measure 1/8″ × 1 1/2″ × 6″ (3.2 mm × 40 mm × 150 mm).

2. Answer the following questions before setting up the welding machine.
   A. Which welding process will you be using, GMAW or FCAW? _____
   B. What electrode diameter will you be using? _____
   C. What is the recommended arc voltage and wire feed speed for the type of metal transfer and electrode diameter you are using?
   Arc voltage: _____
   Wire feed speed: _____
   D. What is the recommended shielding gas and flow rate?
   Shielding gas: _____
   Flow rate: _____
   E. What is the recommended electrode extension? _____
   F. What is the proper work angle? _____
   G. What is the proper travel angle? _____

3. Set up the welding equipment and verify that everything is operating properly.

4.  Align two of the pieces of metal so there is a 1/16″ (1.5 mm) gap between them. Weld three tack welds along the joint. After tacking, the gap should still measure 1/16″ (1.5 mm) along the entire length. Tack weld the third piece onto the first two, as shown in the following drawing. Tack welding can be done in the flat welding position.

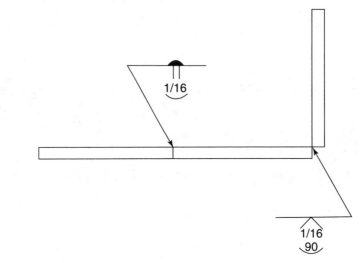

1/16″ = 1.5 mm

5.  Place the metal in a positioner so the weld can be made in the overhead welding position. Weld the butt joint as shown in the drawing. Rotate the metal so both welds are made in the overhead welding position. Fill the V-groove on the outside corner joint from edge-to-edge. The root of the weld pool should look like a keyhole. As you make the weld, keep this keyhole the same size. If you are using a backing, you will not see a keyhole because there is metal behind the joint. You still must melt both pieces evenly.

6.  Tack weld the remaining pieces and weld them. After making each weld, examine it and make any corrections necessary.

## Inspection:

Each butt weld should be straight, slightly convex, and have evenly spaced ripples. Penetration on the back side of the weld should be 100% and should be even along the weld joint. After wire brushing, there should be very little spatter on the plate or on the weld. The outside corner joint should be filled from edge-to-edge and have 100% penetration.

Name: _____  Date: _____

Class: _____  Instructor: _____

Lesson Grade: _____  Instructor's Initials: _____

# Section 4
# Gas Tungsten Arc Welding

## Gas Tungsten Arc Welding Safety Test

### Objectives:

You will be able to discuss the potential safety hazards of GTAW. You will also be able to describe the safety precautions required when working with this equipment.

### Instructions:

Always follow safe practices when welding. If you have safety questions or concerns, ask your instructor. This test does not include questions about every safety topic, but is intended to highlight key items. Review Chapters 19 and 20, especially the material in the Protective Equipment section in Chapter 19. Then, answer or complete the following questions.

_____  1. Which of the following is a primary reason good ventilation must be provided when welding in a closed area?
    A. The fumes generated by GTAW are highly toxic.
    B. The fumes generated by GTAW are very dense and may obscure vision.
    C. Shielding gases may displace air and suffocate the welder.
    D. All of the above.

2. A(n) _____ reduces the pressure in a cylinder to a pressure that is usable for welding.

_____

3. *True or False?* Since GTAW creates a great quantity of fumes, fume extractors or supplied air respirators must always be used.

4. Gas cylinders must be secured to a wall, column, or hand truck with chains or _____ _____.

_____

5. What number filter lens should be placed in your welding helmet for GTAW?

_____

_____  6. Who should connect, inspect, or repair the wires inside a circuit breaker panel?
    A. Each welder.
    B. The welding instructor.
    C. An electrician.
    D. The janitor.

7. *True or False?* Gauntlet gloves must be worn for GTAW.

8. ____ ____ must be worn when using a grinding wheel.

_____

9. Overtightening the regulator nut could ____ the threads.

_____

10. How must the cylinder outlet valve be positioned when you open it to clean it?

_____

_____

## Lesson 19
# GTAW: Equipment and Supplies

## Objectives:
You will be able to describe the gas tungsten arc welding process, the different types of current used, and the equipment used in a GTAW station.

## Instructions:
Read Chapter 19 and study Figures 19-1 through 19-15. Then, answer or complete the following questions.

_____    1. Which of the following currents does *not* provide some cleaning action?
    A. Direct current electrode negative.
    B. Direct current electrode positive.
    C. Alternating current.
    D. DCRP.

2. *True or False?* GTAW is considered to be a very clean process.

3. Label the parts of the GTAW station shown in the following figure.

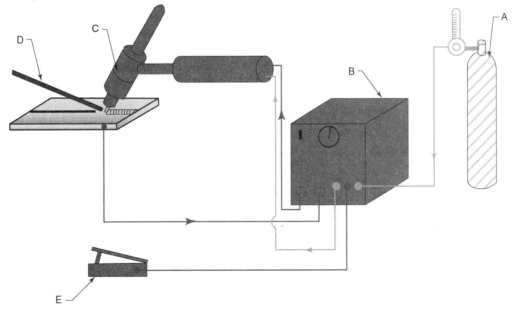

A. _____    D. _____

B. _____    E. _____

C. _____

_____  4. Which of the following shielding gases are used for GTAW?
   A.   Argon and helium.
   B.   Argon and oxygen.
   C.   Helium and carbon dioxide.
   D.   Carbon dioxide and nitrogen.

_____  5. A constant current welding machine allows the welder to vary the current by doing which of the following?
   A.   Changing the shielding gas flow.
   B.   Changing the type of shielding gas.
   C.   Changing the arc length.
   D.   Changing the filler metal.

_____  6. What is the purpose of a gas lens?
   A.   To filter impurities out of the shielding gas.
   B.   To make the shielding gas exit the nozzle in a column.
   C.   To spread shielding gas flow over a larger area.
   D.   To protect the shielding gas from ultraviolet light.

_____  7. Which of the following is *not* a function of a GTAW torch?
   A.   Providing filler metal to the weld pool.
   B.   Cooling the electrode.
   C.   Directing shielding gas.
   D.   Holding the electrode.

_____  8. To permit the welder to vary the current over a wide range while welding, the remote control current switch must be in which position?
   A.   Remote.
   B.   Control panel.
   C.   Neutral.
   D.   None of the above.

9. *True or False?* When current does not flow in the electrode positive half cycle, it is said to be rectified.

10. *True or False?* In GTAW, the tungsten electrode melts and becomes part of the weld.

11. Which type of current flow is illustrated in the following drawing?

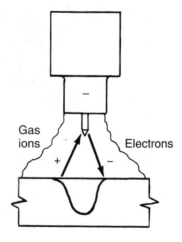

Name _____

12. For GTAW, what is the minimum number shade that should be used in a welding helmet?

_____

_____ 13. Flowmeters measure shielding gas flow rates in which of the following units?
   A. Gallons per hour or liters per hour.
   B. Square feet per minute or square meters per hour.
   C. Cubic feet per hour or liters per minute.
   D. Cubic feet per minute or cubic meters per minute.

14. After the welding current and welding arc have stopped, shielding gas continues to flow. What is this continued flow called?

_____

15. *True or False?* Two current settings are used in pulsed GTAW. These two currents are called forward current and reverse current.

Name: _____     Date: _____

Class: _____     Instructor: _____

Lesson Grade: _____     Instructor's Initials: _____

## Lesson 20
# GTAW: Equipment Assembly and Adjustment

## Objectives:
You will be able to assemble GTAW equipment and set up the welding machine. You will also be able to select the correct current, gas flow, and filler metal.

## Instructions:
Read Chapter 20 and study Figures 20-1 through 20-19. Then, answer or complete the following questions.

1. What tool should be used to tighten a regulator nut to a cylinder?

   _____

2. Each type of tungsten electrode is color-coded for identification. List the color code for these tungsten electrodes.

   EWZr-1: _____

   EWTh-2: _____

   EWP: _____

   EWTh-1: _____

   EWCe-2: _____

   EWLa-1: _____

   EWG: _____

3. For each electrode type listed, indicate whether the electrode can be used with AC only, with DC only, or with both.

   EWZr-1: _____

   EWTh-2: _____

   EWP: _____

   EWTh-1: _____

_____    4.  Which electrode is best for DC welding?
A.  EWP.
B.  EWTh-2.
C.  EWZr-1.
D.  All of the above are equally suitable for DC welding.

_____    5.  When the contactor switch is in the remote position, the current is started by _____.
A.  touching the electrode to the work
B.  pressing a foot pedal or thumb switch
C.  rapidly flipping the switch back and forth
D.  The current cannot be started when the contactor switch is in the remote position.

*For questions 6–10, refer to the following chart to answer questions about welding a lap joint on aluminum 3/16" (4.8 mm) thick.*

| Aluminum and Aluminum Alloys—High-Frequency AC | | | | | Gas | | |
|---|---|---|---|---|---|---|---|
| Metal thickness | Joint type | Tungsten electrode diameter | Welding rod diameter (if req'd.) | Amperage | Type | Flow (cfh) | L/min |
| 1/16" (1.6 mm) | Butt | 1/16" (1.6 mm) | 1/16" (1.6 mm) | 60–85 | Argon | 15 | 7.0 |
|  | Lap | 1/16" | 1/16" | 70–90 | Argon | 15 | 7.0 |
|  | Corner | 1/16" | 1/16" | 60–85 | Argon | 15 | 7.0 |
|  | Fillet | 1/16" | 1/16" | 75–100 | Argon | 15 | 7.0 |
| 1/8" (3.2 mm) | Butt | 3/32"–1/8" (2.4 mm–3.2 mm) | 3/32" (2.4 mm) | 125–150 | Argon | 20 | 9.0 |
|  | Lap | 3/32"–1/8" | 3/32" | 130–160 | Argon | 20 | 9.0 |
|  | Corner | 3/32"–1/8" | 3/32" | 120–140 | Argon | 20 | 9.0 |
|  | Fillet | 3/32"–1/8" | 3/32" | 130–160 | Argon | 20 | 9.0 |
| 3/16" (4.8 mm) | Butt | 1/8"–5/32" (3.2 mm–4.0 mm) | 1/8" (3.2 mm) | 180–225 | Argon | 20 | 9.0 |
|  | Lap | 1/8"–5/32" | 1/8" | 190–240 | Argon | 20 | 9.0 |
|  | Corner | 1/8"–5/32" | 1/8" | 180–225 | Argon | 20 | 9.0 |
|  | Fillet | 1/8"–5/32" | 1/8" | 190–240 | Argon | 20 | 9.0 |
| 1/4" (6.4 mm) | Butt | 5/32"–3/16" (4.0 mm–4.8 mm) | 3/16" (4.8 mm) | 240–280 | Argon | 25 | 12.0 |
|  | Lap | 5/32"–3/16" | 3/16" | 250–320 | Argon | 25 | 12.0 |
|  | Corner | 5/32"–3/16" | 3/16" | 240–280 | Argon | 25 | 12.0 |
|  | Fillet | 5/32"–3/16" | 3/16" | 250–320 | Argon | 25 | 12.0 |

6.  What diameter electrode should be used?

_____

7.  What diameter filler rod should be used?

_____

8.  What is the current range?

_____

9.  What type of shielding gas is recommended?

_____

10.  What is the recommended shielding gas flow rate?

_____

Name _____

_____ 11. Which type of electrodes *cannot* be used with both AC and DC welding?
A. Pure tungsten electrodes.
B. Thoria electrodes.
C. Zirconia electrodes.
D. All of the above.

_____ 12. When grinding an electrode, the grind marks should be ____.
A. lengthwise
B. circular
C. around the electrode
D. lengthwise for DC welding and around the electrode for AC welding

_____ 13. Which of the following may be connected to the remote control connection on the welding machine?
A. Foot pedal.
B. Torch.
C. Shielding gas.
D. Cooling water.

14. Match the type of welding current from the list at right that is best for welding the base metals listed at left.

_____ Nickel.

_____ Aluminum.

_____ Low-carbon steel.

_____ Stainless steel.

_____ Titanium.

A. DCEP (DCRP).
B. DCEN (DCSP).
C. AC.

15. *True or False?* You must grind or form a new ball on an electrode whenever you use a new electrode and whenever the electrode becomes contaminated.

16. Name the two parts of the torch that must be changed whenever the electrode diameter is changed.

_____

_____

_____ 17. Which of the following statements is true for DC welding?
A. High-frequency voltage must be used to start the arc.
B. High-frequency voltage may be used to start the arc.
C. High-frequency voltage must be used continuously.
D. High-frequency voltage must never be used.

18. *True or False?* For AC welding, the electrode tip is formed into a ball.

## Assigned Job 20-1
# Selecting and Preparing an Electrode

## Objectives:

In this job, you will learn to select the proper electrode for a welding job and to properly prepare an electrode for DC welding. Read the Selecting and Preparing the Electrode section and study Figures 20-13 through 20-19 in Chapter 20 of the text.

> **Note**
>
> Do not attempt this job until you have read all safety precautions, satisfactorily completed the *Gas Tungsten Arc Welding (GTAW) Safety Test*, and been approved by your instructor.

1. For the following base metals, place check marks to indicate which welding current, AC or DC, is best for GTAW. Also indicate which electrode, a thoria tungsten electrode (EWTh-2) or a pure tungsten electrode (EWP), is best for welding each base metal.

| Base Metal | AC | DC | EWTh–2 | EWP |
|---|---|---|---|---|
| High-carbon steel | | | | |
| Cast iron | | | | |
| Aluminum | | | | |
| Magnesium | | | | |
| Stainless steel | | | | |

2. The proper electrode must be selected to match the required base metal and thickness to be welded. Which of the following electrodes would you choose for each of the welding jobs described? Refer to Figure 20-15 in the text. Use each answer only once.

_____ AC welding aluminum using 70 amps.

_____ AC welding magnesium using 165 amps.

_____ DCEN welding stainless steel using 80 amps.

_____ DCEN welding low alloy steel using 175 amps.

A. 2% thoria tungsten 1/16″ (1.6 mm) in diameter.

B. Pure tungsten 1/16″ (1.6 mm) in diameter.

C. Pure tungsten 1/8″ (3.2 mm) in diameter.

D. 2% thoria tungsten 3/32″ (2.4 mm) in diameter.

3. Obtain one 1/16″ (1.6 mm) tungsten electrode. Complete the following steps to grind a point on the electrode.

4. You should always wear safety glasses when grinding. However, gloves are usually not worn when grinding tungsten electrodes because the task requires good dexterity.

5. Stand facing the grinder and hold the electrode in line with the rotating wheel, but angled toward the top or bottom of the grinding wheel. Carefully bring the electrode into contact with the wheel. If you have the electrode properly positioned, the grinding wheel will create lengthwise grinding marks on it. Remember, widthwise grinding marks are unacceptable. See Figure 20-16 in the text.

6. As the grinding wheel starts to grind an angle on the electrode, rotate the electrode between your fingers. As grinding continues, this simple action will create a cone-shaped taper at the end of the electrode that is properly centered on the electrode's diameter.

7. Continue rotating and grinding the electrode until a taper length of two to three times the electrode's diameter is achieved. For a 1/16" (1.6 mm) electrode, the taper length will be 1/8" to 3/16" (3.0 mm to 5.0 mm).

8. Once a tapered end is ground on the electrode, the tip must be blunted. Hold the electrode square to the grinding wheel so only the tip will touch the grinding wheel. Lightly touch the end of the electrode to the grinding wheel. This removes the tip of the electrode. Examine the end of the electrode. The flat tip on the end should be 1/4 of the electrode diameter. For a 1/16" (1.6 mm) electrode, the flat spot would be 1/64" (0.4 mm).

9. Show your instructor your prepared electrode.

## Inspection:

A properly ground electrode for DC welding will have the grinding lines running lengthwise on the electrode. The length of the taper for most applications will be two to three times the electrode diameter. The tip of the electrode will be blunted slightly.

# Assigned Job 20-2
# Assembling the Torch

## Objective:

In this job, you will learn the parts of a GTAW torch and how to assemble it.

> **Note**
> Do not attempt this job until you have read all safety precautions, satisfactorily completed the *Gas Tungsten Arc Welding (GTAW) Safety Test*, and been approved by your instructor.

1. You will need a complete torch, including an electrode. This lesson assumes you are beginning with a completely assembled torch.

2. Label the parts of the torch shown in the following photograph.

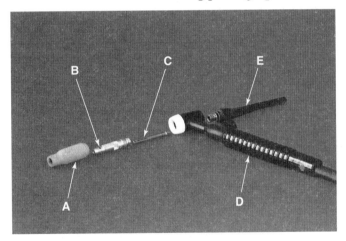

A. _____     D. _____

B. _____     E. _____

C. _____

3. Loosen the end cap and remove the electrode.

4. Remove the end cap. Place your hand over the hole where the end cap was removed. Turn the torch upside down and the collet will fall out.

5. Remove the nozzle.

6. Using a properly fitting wrench, remove the collet body. This completes the torch disassembly.

7. Hold the collet body in your hand. Place the collet into the collet body and the electrode into the collet. Notice how well they fit. When these parts are inside the torch and the end cap is tightened, the slits in the collet close down onto the electrode. This holds the electrode in place.

8. Begin reassembly of the torch by installing the collet body. Using a properly fitting wrench, secure the collet body. Do not overtighten it. Screw on the nozzle. Place the collet into the top of the torch. Screw the end cap on, but do not tighten it completely.

9. Install the electrode so it extends 1/8" to 3/16" (3.0 mm to 5.0 mm) beyond the nozzle. Tighten the end cap.

10. Review the following points:
    A. The electrode must be removed for grinding or when switching types of electrodes.
    B. The end cap is loosened to change the distance that the electrode extends.
    C. The collet and collet body do not need to be changed unless the diameter of the electrode is changed.
    D. The nozzle may be changed depending on the work and joint configuration to be welded.

## Inspection:

You should be able to disassemble and reassemble a GTAW torch. You should know how to remove and replace the electrode, the collet and collet body, and the nozzle. Your instructor may ask you to demonstrate one or more of these skills.

## Lesson 21
# GTAW: Flat Welding Position

## Objectives:

You will be able to discuss how acceptable welds are made using gas tungsten arc welding in the flat welding position. You will also be able to demonstrate GTAW.

## Instructions:

Read Chapter 21 and study Figures 21-1 through 21-19. Then, answer or complete the following questions.

1. What is the proper travel angle for laying a weld bead on plate?

   _____

_____  2. When using high-frequency voltage to start an arc, how far should the end of the electrode be held from the surface of the work?
   A. 1/8″ (3.0 mm).
   B. 1/4″ (6.0 mm).
   C. 1/2″ (13 mm).
   D. 1″ (25 mm).

_____  3. Which of the following problems is caused by using a filler rod that is too small?
   A. Large bumps in the weld bead.
   B. The rod is used up quickly.
   C. Excessive penetration.
   D. All of the above.

4. *True or False?* To obtain more current while welding, move the electrode farther from the work.

5. What is the minimum number filter lens suggested for use in the helmet for GTAW?

   _____

6. *True or False?* Hot shortness is a condition in which a metal has very little strength when it is very hot. This can cause the molten pool to fall through during welding.

*For Questions 7–10, refer to following chart. Answer the following questions regarding welding a fillet weld on 1/8" (3.2 mm) thick mild steel.*

| Mild Steel – DCEN (DCSP) | | | | | | | |
|---|---|---|---|---|---|---|---|
| Metal thickness | Joint type | Tungsten electrode diameter | Welding rod diameter (if req'd.) | Amperage | Gas | | |
| | | | | | Type | Flow (cfh) | L/min* |
| 1/16" (1.6 mm) | Butt | 1/16" (1.6 mm) | 1/16" (1.6 mm) | 60–70 | Argon | 15 | 7.0 |
| | Lap | 1/16" | 1/16" | 70–90 | Argon | 15 | 7.0 |
| | Corner | 1/16" | 1/16" | 60–70 | Argon | 15 | 7.0 |
| | Fillet | 1/16" | 1/16" | 70–90 | Argon | 15 | 7.0 |
| 1/8" (3.2 mm) | Butt | 1/16"–3/32" (1.6 mm–2.4 mm) | 3/32" (2.4 mm) | 80–100 | Argon | 15 | 7.0 |
| | Lap | 1/16"–3/32" | 3/32" | 90–115 | Argon | 15 | 7.0 |
| | Corner | 1/16"–3/32" | 3/32" | 80–100 | Argon | 15 | 7.0 |
| | Fillet | 1/16"–3/32" | 3/32" | 90–115 | Argon | 15 | 7.0 |
| 3/16" (4.8 mm) | Butt | 3/32" (2.4 mm) | 1/8" (3.2 mm) | 115–135 | Argon | 20 | 9.0 |
| | Lap | 3/32" | 1/8" | 140–165 | Argon | 20 | 9.0 |
| | Corner | 3/32" | 1/8" | 115–135 | Argon | 20 | 9.0 |
| | Fillet | 3/32" | 1/8" | 140–170 | Argon | 20 | 9.0 |
| 1/4" (6.4 mm) | Butt | 1/8" (3.2 mm) | 5/32" (4.0 mm) | 160–175 | Argon | 20 | 9.0 |
| | Lap | 1/8" | 5/32" | 170–200 | Argon | 20 | 9.0 |
| | Corner | 1/8" | 5/32" | 160–175 | Argon | 20 | 9.0 |
| | Fillet | 1/8" | 5/32" | 175–210 | Argon | 20 | 9.0 |

*Liters per minute

7. What diameter electrode should be used?

_____

8. What diameter filler rod should be used?

_____

9. What is the recommended current?

_____

10. What type of shielding gas is recommended?

_____

_____     11.    How far should the tip of the electrode be held from the work after the welding arc is stabilized?
A.   Less than 1/16" (1.5 mm).
B.   1/16"–1/8" (1.5 mm–3.0 mm).
C.   1/8"–3/16" (3.0 mm–5.0 mm).
D.   The distance depends on the material thickness.

12. *True or False?* A short welding arc length can cause porosity.

13. List two ways to strike an arc.

_____

_____

Name _____

14. Refer to the following chart to select a proper filler metal electrode to weld each of the base metals listed at left.

| Base Metal | Recommended Filler Metal |
|---|---|
| Aluminum | ER1100, ER4043, ER5356 |
| Copper and copper alloys | ERCu, ERCuSi-A, ERCuAl-A1 |
| Magnesium | ERAZ61A, ERAZ92A |
| Nickel and nickel alloys | ERNi-1, ERNiCr-3, ERNiCrMo-3 |
| Carbon steel | ER70S-2, ER70S-6 |
| Low alloy steel | ER80S-B2, ER80S-D2 |
| Stainless steel | ER308, ER308L, ER316, ER347 |
| Titanium | ERTi-1 |
| Zirconium | ERZr-2 |

_____  Aluminum.          A.  ERAZ61A.

_____  Carbon steel.       B.  ERTi-1.

_____  Stainless steel.     C.  ER308.

_____  Magnesium.          D.  ER5356.

_____  Titanium.            E.  ER70S-2.

15. Indicate the correct angles for welding a butt joint in the flat welding position.

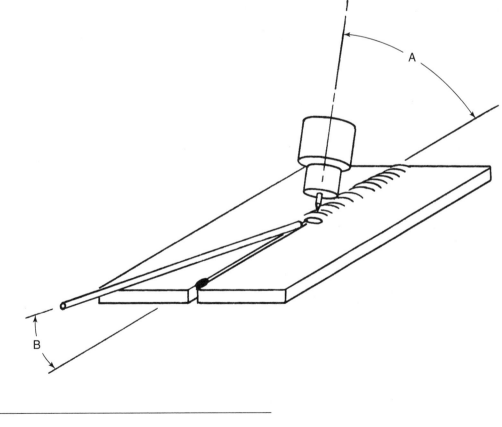

A. _____

B. _____

16. *True or False?* As soon as the arc is stable, you can begin to move the torch forward.

17. The welding rod must be kept very close to the weld pool and the welding arc so it remains in the area covered by _____ _____.

---

**A note to students about the Assigned Jobs in Section 4:**

Gas tungsten arc welding is a very clean process. It is used in industry to weld many different metals, such as mild steel, stainless steel, aluminum, and titanium. You should practice on as many of these as possible.

To allow you to learn the skills required for GTAW, the jobs in this section are organized from the easiest to the hardest. You will weld on mild steel while learning the basic skills (Assigned Job 21-1 through Assigned Job 21-6). After learning the basics, you then will learn to run beads on stainless steel and on aluminum. Assigned Jobs 21-2, 21-3, and 21-5 can be repeated using stainless steel and aluminum.

Assigned Jobs 21-9 through 22-9 do not specify a metal to weld. Your instructor will decide which type of metal will be used. These jobs can be repeated with different metals.

## Assigned Job 21-1
# Creating a Continuous Weld Pool

## Objective:

In this job, you will learn to set up a GTAW machine. You will strike an arc and move a molten weld pool along a plate. You will learn to control the weld pool and properly stop the arc.

> **Note**
> Do not attempt this job until you have read all safety precautions, satisfactorily completed the *Gas Tungsten Arc Welding (GTAW) Safety Test*, and been approved by your instructor.

1. Obtain three pieces of low-carbon steel that measure $1/16'' \times 4'' \times 6''$ (1.6 mm × 100 mm × 150 mm).

2. Answer the following questions about welding $1/16''$ (1.6 mm) mild steel. Refer to the Selecting the Weld Current and Selecting and Preparing the Electrode sections in Chapter 20. Also refer to Figures 20-8 and 20-13 in the text.
   A. What type of current is used? _____
   B. What type of electrode is used? _____
   C. What diameter of electrode is used? _____
   D. What amperage is used? (Use the amperage recommended for a butt weld.)_____
   E. What type of shielding gas is used? _____
   F. What is the recommended shielding gas flow rate?_____

3. Set up the welding machine using the following checklist. Refer to the Welding Machine Settings section in Chapter 20.
   - Set the type of electrical current: AC, DCEN (DCSP), or DCEP (DCRP). (See your answer to question 2A.)
   - Set the current range.
   - Set the desired current using the current control knob. (See your answer to question 2D.)
   - Set the high-frequency control for start only.
   - Set the contactor switch to the remote position.
   - Set the current switch to the panel position.
   - Set the post flow for about 10 seconds.

4. Obtain the correct electrode type and diameter. Prepare the electrode and install it in the torch. (See your answers to questions 2B and 2C.)

5. Adjust the shielding gas flow rate for the proper amount. (See your answer to question 2F.) Refer to the Adjusting the Shielding Gas Flowmeter section in Chapter 20.

6. Clean the base metal surface with a wire brush. Remove all rust, grease, and oil. Next, use an awl or other sharp object to scratch four straight lines into the metal. You just need to scratch the surface, not gouge a deep, wide groove. Soapstone is not used because it can contaminate the weld process. The lines should be about 1" (25 mm) apart and 1/2" (13 mm) from the edge. The lines should run along the 6" (150 mm) side of the metal, as shown in the following drawing. Place the metal in front of you on the work table. Make sure you can move the torch freely from one side to the other.

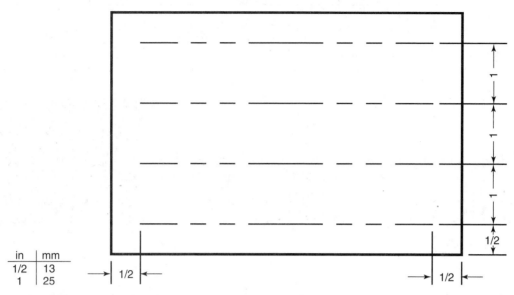

| in | mm |
|----|----|
| 1/2 | 13 |
| 1 | 25 |

7. Answer the following questions to prepare yourself for laying the weld beads:
   A. What is the proper work angle? _____
   B. What is the proper travel angle? _____

8. Place the foot pedal on the floor where you can reach it comfortably.

9. Make sure you have the correct lens in your helmet and are wearing the proper clothes.

10. Hold the torch about 1/8″ (3.0 mm) above the work in a forehand welding position. Hold the torch at the right end of one of the lines on the plate if you are right-handed, or at the left end if you are left-handed.

11. Lower your helmet and press the foot pedal. The high-frequency voltage will jump the gap and the welding arc will start.

12. After the welding arc starts, move the electrode so it is 1/16″–1/8″ (1.5 mm–3.0 mm) from the work.

13. Hold the electrode in the same place until a molten weld pool forms. The weld pool should be about 1/8″–3/16″ (3.0 mm–5.0 mm) wide. The weld pool may start to sag as it reaches the right size.

14. As soon as the weld pool becomes the right size, begin moving it forward. You must move fairly slowly. As you move the weld pool, keep it the same size.
    • Keep the electrode the correct distance from the work. If you dip the electrode into the molten weld pool, stop welding (see step 15). If you dip the electrode into the molten weld pool or if the welding arc is jumping off the electrode in uncontrollable directions, the electrode has been contaminated and must be reground. Remove the electrode from the torch. Grind the electrode to remove all contamination. A clean tip is always used for welding. Reinstall the electrode into the torch and start welding again.
    • If you need to change the current slightly, you can move the electrode closer to the work for more current or farther away for less current.
    • To significantly change the current, stop welding and change the setting on the front of the welding machine.

15. When you reach the end of a weld or must stop for any reason, take your foot off the pedal. The welding arc will stop. Remember, shielding gas continues to flow even though the welding arc is stopped. Keep the nozzle over the end of the weld until the gas stops flowing.

16. Inspect the weld bead as described in the following section. Create four weld beads on each of the pieces of metal. Your last piece should have four good-quality weld beads on it.

## Inspection:

Each weld bead should be straight. The ripples should be evenly spaced and the width should remain the same. There should be no evidence of dipping the tungsten electrode into the weld pool or of porosity due to a very long arc length.

Name: _____     Date: _____

Class: _____     Instructor: _____

Lesson Grade: _____     Instructor's Initials: _____

## Assigned Job 21-2
# Welding a Lap Joint without Filler Metal

## Objective:

In this job, you will learn to strike an arc and move a molten weld pool along a lap joint. You will learn how to control the pool and how to properly stop the arc.

> **Note**
> Do not attempt this job until you have read all safety precautions, satisfactorily completed the *Gas Tungsten Arc Welding (GTAW) Safety Test*, and been approved by your instructor.

1. Obtain six pieces of low-carbon steel that measure 1/16″ × 1 1/2″ × 6″ (1.6 mm × 40 mm × 150 mm).

2. Answer the following questions about welding 1/16″ (1.6 mm) mild steel. Refer to the Selecting the Weld Current and Selecting and Preparing the Electrode sections and Figures 20-8 and 20-13 in Chapter 20 of the text.

   A. What type of current is used? _____

   B. What type of electrode is used? _____

   C. What diameter of electrode is used? _____

   D. What amperage is used to weld a lap joint? _____

   E. What type of shielding gas is used? _____

   F. What is the recommended shielding gas flow rate? _____

3. Set up the welding machine using the following checklist. Refer to the Welding Machine Settings section in Chapter 20 of the text.
   - Set the type of electrical current: AC, DCEN (DCSP), or DCEP (DCRP). (See your answer to question 2A.)
   - Set the current range.
   - Set the desired current using the current control knob. (See your answer to question 2D.)
   - Set the high-frequency control for start only.
   - Set the contactor switch to the remote position.
   - Set the current switch to the panel position.
   - Set the post flow for about 10 seconds.

4. Obtain one tungsten electrode of the correct type and diameter. Prepare the electrode and install it in the torch. (See your answers to questions 2B and 2C.)

5. Adjust the shielding gas flow for the proper rate. (See your answer to question 2F.)

6. Clean the metal to be welded. Align and tack weld two pieces of metal to form a lap joint, as shown in the following drawing. Place the metal in front of you on the work table. Support the metal so the weld joint is in the flat welding position.

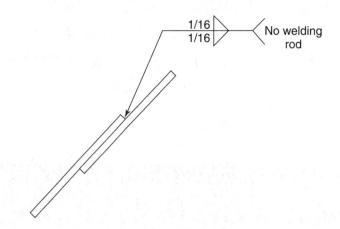

1/16″ = 1.6 mm

7. Answer the following questions to prepare yourself for welding the joint:
   A. What is the proper work angle? _____
   B. What is the proper travel angle? _____

8. Hold the torch over one end of the joint, so you will be welding forehand. Hold the electrode over the centerline of the joint.

9. Press on the foot pedal and strike an arc. After the welding arc starts, move the electrode so it is 1/16″–1/8″ (1.5 mm–3.0 mm) from the work.

10. Hold the electrode in the same place until a molten weld pool forms. You will melt the edge to create a fillet weld. Form a C-shaped weld pool. You may need to point the torch more toward the surface than toward the edge.

11. As soon as the weld pool reaches the right size, begin moving forward. Keep the pool the same size as you move along the joint. The fillet weld should have equal legs. See the drawing in step 6 for the correct size.
    • Do not dip the electrode into the weld. If you do, stop welding and grind the electrode to remove any contamination. Reinstall the electrode and start welding again.

12. When you reach the end of a weld, take your foot off the foot pedal or release the thumb switch. The welding arc will stop. Keep the nozzle over the end of the weld until the shielding gas stops flowing.

13. Inspect the weld as described in the following section. Weld a fillet weld on the opposite side and inspect it. Make any changes in current on the welding machine. Repeat this job with the remaining material. Your last piece should have two good-quality fillet welds on it.

## Inspection:

Each weld bead should be straight. The edge should be melted a consistent amount. The fillet weld should flow and fuse smoothly into both pieces. There should be no overlap. The weld face should be flat or slightly concave. There should be no evidence of dipping the tungsten electrode into the weld pool or of porosity due to a very long welding arc length.

## Assigned Job 21-3
# Welding a Square-Groove Butt Joint on an Outside Corner Joint without Filler Metal

## Objective:

In this job, you will learn to weld an outside corner joint in the flat position. You will strike an arc, learn to control the pool, and properly stop the arc. No filler metal will be added.

---

**Note**

Do not attempt this job until you have read all safety precautions, satisfactorily completed the *Gas Tungsten Arc Welding (GTAW) Safety Test*, and been approved by your instructor.

---

1. Obtain eight pieces of low-carbon steel that measure $1/16'' \times 1\,1/2'' \times 6''$ (1.6 mm $\times$ 40 mm $\times$ 150 mm).

2. Answer the following questions about welding $1/16''$ (1.6 mm) mild steel. Refer to the Selecting the Welding Current and Selecting and Preparing the Electrode and Figures 20-8 and 20-13 in Chapter 20 of the text.
   A. What type of current is used? _____
   B. What type of electrode is used? _____
   C. What diameter of electrode is used? _____
   D. What amperage is used to weld a corner joint? _____
   E. What type of shielding gas is used? _____
   F. What is the recommended shielding gas flow rate? _____

3. Set up the welding machine using the following checklist. Refer to the Welding Machine Settings section in Chapter 20.
   - Set the type of electrical current: AC, DCEN (DCSP), or DCEP (DCRP). (See your answer to question 2A.)
   - Set the current range.
   - Set the desired current using the current control knob. (See your answer to question 2D.)
   - Set the high-frequency control for start only.
   - Set the contactor switch to the remote position.
   - Set the current switch to the panel position.
   - Set the post flow for about 10 seconds.

4. Obtain a tungsten electrode of the correct type and diameter. Prepare the electrode and install it in the torch. (See your answers to questions 2B and 2C.)

5. Adjust the shielding gas flow for the proper rate. (See your answer to question 2F.)

6. Clean the metal to be welded. Align and tack weld two pieces of metal to form an outside corner joint, as shown in the following drawing. Place the metal in front of you on the work table. Support the metal so the weld joint is in the flat welding position.

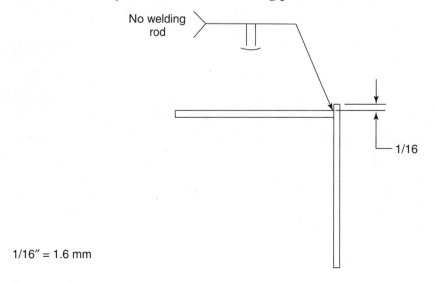

1/16″ = 1.6 mm

7. Answer the following questions to prepare yourself for welding the joint:
   A. What is the proper work angle? _____
   B. What is the proper travel angle? _____

8. Hold the torch over one end of the joint with the electrode positioned over the weld axis, so you can weld the joint with the forehand technique.

9. Press on the foot pedal and strike an arc. After the welding arc starts, move the electrode so it is 1/16″–1/8″ (1.5 mm–3.0 mm) from the work.

10. The overlapping metal will melt and begin to form a weld pool. Point the torch toward the surface of the overlapped piece, so it also melts. When the weld pool flows and joins both pieces of metal, you can begin to move forward. The weld pool should be about 1/8″–3/16″ (3.0 mm–5.0 mm) wide. Continue to melt the overlap material, using the metal to join the two pieces together. By moving the torch onto the overlapping piece and changing the angle slightly, you will be able to put a radius on the corner.

11. Weld the outside corner joint. Be careful not to dip your tungsten electrode into the weld. If you do, regrind the tip of the electrode before continuing to weld.

12. Inspect the weld. Make any necessary changes in current on the welding machine. Repeat this job three times with the remaining material. Show these three weldments to your instructor for a grade.

## Inspection:

The weld should flow and join both pieces. The ripples should be evenly spaced and be the same width along the entire weld joint. There should be no visible defects.

Name: _____   Date: _____

Class: _____   Instructor: _____

Lesson Grade: _____   Instructor's Initials: _____

## Assigned Job 21-4
# Making a Weave Bead on Plate

## Objectives:

In this job, you will learn to develop a wide weld pool by weaving, or moving, the torch from side to side. You will strike an arc and move a wide molten weld pool along a plate. You will learn to control the weld pool and properly stop the welding arc.

> **Note**
> Do not attempt this job until you have read all safety precautions, satisfactorily completed the *Gas Tungsten Arc Welding (GTAW) Safety Test*, and been approved by your instructor.

1. Obtain three pieces of low-carbon steel that measure 1/8″ × 4″ × 6″ (3.2 mm × 100 mm × 150 mm).

2. Answer the following questions about welding 1/8″ (3.2 mm) mild steel. Refer to the Selecting the Weld Current and Selecting and Preparing the Electrode sections and Figures 20-8 and 20-13 in Chapter 20 of the text.
   A. What type of current is used? _____
   B. What type of electrode is used? _____
   C. What diameter of electrode is used? _____
   D. What amperage is used? Use the amperage recommended for a butt weld. _____
   E. What type of shielding gas is used? _____
   F. What is the recommended shielding gas flow rate? _____

3. Set up the welding machine using the following checklist. Refer to the Welding Machine Settings section in Chapter 20.
   - Set the type of electrical current: AC, DCEN (DCSP), or DCEP (DCRP). (See your answer to question 2A.)
   - Set the current range.
   - Set the desired current using the current control knob. (See your answer to question 2D.)
   - Set the high-frequency control for start only.
   - Set the contactor switch to the remote position.
   - Set the current switch to the panel position.
   - Set the post flow for about 10 seconds.

4. Obtain one tungsten electrode of the correct type and diameter. Prepare the electrode and install it in the torch. (See your answers to questions 2B and 2C.)

5. Adjust the shielding gas flow for the proper rate. Refer to the Adjusting the Shielding Gas Flowmeter section in Chapter 20. (See your answer to question 2F.)

6. Clean the base metal surface with a wire brush. Remove all rust, grease, and oil. Use an awl or

similar sharp object and scratch four straight lines into the metal. The lines should be about 1″ (25 mm) apart and run along the 6″ (150 mm) side of the metal. See the drawing in Assigned Job 21-1. Place the metal in front of you on the worktable. Make sure you can move the torch freely from one side to the other.

7. Answer the following questions to prepare yourself for creating the weld pool:
   A. What is the proper work angle? _____
   B. What is the proper travel angle? _____

8. Hold the torch in a forehand welding position at the end of one of the lines. Strike an arc.

9. After the welding arc stabilizes, begin to move the torch from side to side to create a weld pool that is about 3/8″–1/2″ (9.5 mm–13 mm) wide. You may use a circular motion instead of just a side-to-side movement.

10. When the weld pool grows to the right size, begin moving forward. As you move forward, continue to move the torch in a circular motion or a side-to-side motion. Keep the weld pool the same width as you move along the plate.

11. Observe the following rules as you complete the weld:
    • Keep the electrode the correct distance from the work.
    • Stop the welding arc when you are 1/2″ (13 mm) from the far edge of the metal.
    • Make any large current changes on the welding machine.
    • Make small current changes by moving the electrode closer to or farther from the work.
    • Be careful not to dip the electrode into the weld.

12. Inspect the weld bead and make any needed current changes. Lay four weave beads on each of the pieces of metal you have. Your last piece should have four good-quality weave beads on it.

## Inspection:

A weld bead made with a weaving motion should look the same as a weld bead made with a straight motion, except that it should be wider. The weld bead should be the same width along its entire length. The ripples should be evenly spaced. There should be no evidence of defects.

Name: _____   Date: _____

Class: _____   Instructor: _____

Lesson Grade: _____   Instructor's Initials: _____

## Assigned Job 21-5
# Welding a Square-Groove Edge Joint without Filler Metal

## Objective:

In this job, you will learn to control a wide weld pool by using a weaving motion while welding an edge joint.

> **Note**
> Do not attempt this job until you have read all safety precautions, satisfactorily completed the *Gas Tungsten Arc Welding (GTAW) Safety Test*, and been approved by your instructor.

1.  Obtain four pieces of low-carbon steel that measure 1/8″ × 1 1/2″ × 6″ (3.2 mm × 40 mm × 150 mm).

2.  Answer the following questions about welding 1/8″ (3.2 mm) mild steel. Refer to the Selecting the Welding Current and Selecting and Preparing the Electrode sections and Figures 20-8 and 20-13 in Chapter 20 of the text.

    A.  What type of current is used? _____

    B.  What type of electrode is used? _____

    C.  What diameter of electrode is used? _____

    D.  What amperage is used? (Use the amperage recommended for a lap joint.) _____

    E.  What type of shielding gas is used? _____

    F.  What is the recommended shielding gas flow rate? _____

3.  Set up the welding machine using the following checklist. Refer to the Welding Machine Settings section in Chapter 20 of the text.
    - Set the type of electrical current: AC, DCEN (DCSP), or DCEP (DCRP). (See your answer to question 2A.)
    - Set the current range.
    - Set the desired current using the current control knob. (See your answer to question 2D.)
    - Set the high-frequency control for start only.
    - Set the contactor switch to the remote position.
    - Set the current switch to the panel position.
    - Set the post flow for about 10 seconds.

4.  Obtain a tungsten electrode of the correct type and diameter. Prepare the electrode and install it in the torch. (See your answers to questions 2B and 2C.)

5.  Adjust the shielding gas flow for the proper rate. (See your answer to question 2F.)

6. Clean the metal to be welded. Align and tack weld two pieces of metal to form an edge joint, as shown in the following drawing. Place the metal in front of you on the work table. Support the metal so the weld joint is in the flat welding position.

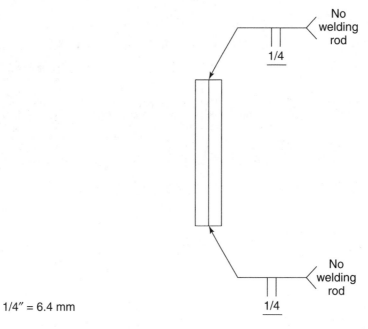

7. Answer the following questions to prepare yourself for welding the joint:
   A. What is the proper work angle? _____
   B. What is the proper travel angle? _____

8. Hold the torch over one end of the joint so you will be welding forehand. Position the electrode over the weld axis.

9. Strike an arc and create a weld pool that is as wide as the two pieces of metal being welded. As soon as the weld pool is the right size, begin moving it forward. Keep the weld pool the same size as you move it along the joint. Do not let the weld become too wide and run over the edges.

10. Inspect the weld as directed in the following section. Make any necessary changes in current on the welding machine. Weld the edge joint on the opposite edge and inspect it. Repeat this job with the remaining material. Your last piece should have a good-quality weld on both edges.

## Inspection:

Each weld bead should be as wide as the edge joint. The weld should not sag over the edge. There should be no evidence of dipping the tungsten electrode into the weld pool or of porosity due to a very long welding arc length.

Name: _____  Date: _____

Class: _____  Instructor: _____

Lesson Grade: _____  Instructor's Initials: _____

## Assigned Job 21-6
# Making Stringer and Weave Beads on Plate with Filler Metal

## Objectives:

In this job, you will learn to gas tungsten arc weld with filler metal. You will strike the welding arc and move a molten weld pool along the plate. You will learn to control the weld pool and to properly stop the arc and fill the weld pool. You will also learn to vary the welding current while welding.

> **Note**
> Do not attempt this job until you have read all safety precautions, satisfactorily completed the *Gas Tungsten Arc Welding (GTAW) Safety Test*, and been approved by your instructor.

1. Obtain eight pieces of low-carbon steel that measure 1/8″ × 4″ × 6″ (3.2 mm × 100 mm × 150 mm).

2. Answer the following questions about welding 1/8″ (3.2 mm) mild steel. Refer to the Selecting the Weld Current and Selecting and Preparing the Electrode sections and Figures 20-8 and 20-13 in Chapter 20 of the text.

   A. What type of current is used? _____

   B. What type of electrode is used? _____

   C. What diameter of electrode is used? _____

   D. What diameter of filler metal is used? _____

   E. What amperage is used? (Use the amperage recommended for a butt weld.) _____

   F. What type of shielding gas is used? _____

   G. What is the recommended shielding gas flow rate? _____

3. Set up the welding machine using the following checklist. Refer to the Welding Machine Settings section in Chapter 20 of the text.
   - Set the type of electrical current: AC, DCEN (DCSP), or DCEP (DCRP). (See your answer to question 2A.)
   - Set the current range.
   - Set the desired current using the current control knob.
   - Set the high-frequency control for start only.
   - Set the contactor switch to the remote position.
   - Set the current switch to the panel position.
   - Set the post flow for about 10 seconds.

4. Obtain the correct electrode. Prepare the electrode and install it in the torch. (See your answers to questions 2B and 2C.)

5. Obtain 12 pieces of the correct diameter filler metal. (See your answer to question 2D.)

6. Adjust the shielding gas flow for the proper rate. (See your answer to question 2G.)

7. Clean the metal to be welded. Use an awl or other sharp object and lightly scratch four straight lines into the metal. The lines should be about 1″ (25 mm) apart and 1/2″ (13 mm) from the edge. The lines should run the 6″ (150 mm) length of the metal, as shown in the following drawing. Place the metal in front of you on the work table.

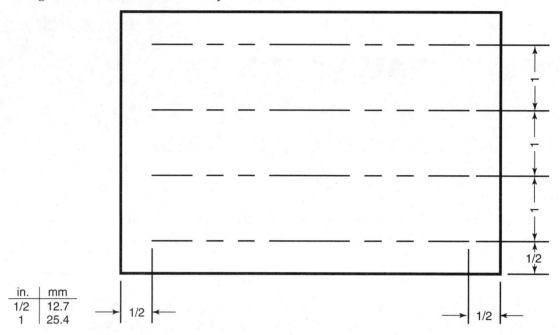

| in. | mm |
|-----|------|
| 1/2 | 12.7 |
| 1 | 25.4 |

8. Strike an arc. After the welding arc starts, move the electrode so it is 1/16″–1/8″ (1.5 mm–3.0 mm) from the work.

9. Hold the electrode in the same place until a molten weld pool forms. The pool should be about 3/16″–1/4″ (4.5 mm–6.0 mm) wide. While the weld pool is forming, bring the filler metal close to it. Hold the filler metal at a 15°–20° angle.

10. As soon as the weld pool reaches the right size, add the filler metal to its front edge. Refer to the Welding a Bead on Plate with Filler Metal section in Chapter 21 of the text. After adding filler metal, begin moving the pool forward. Keep the filler metal close to the weld pool. Keep the weld pool the same size as you move it forward. Add filler metal as required to make a slightly convex weld bead.
    • Do not dip the electrode into the weld pool or touch the filler metal to the electrode. If you do, stop welding and grind the electrode to remove any contamination. Reinstall the electrode and start welding again.

11. When you reach the end of a weld, you must fill the weld pool with filler metal. To do this, move the torch to the rear of the weld pool. Add filler metal to the pool. Move the torch forward so it smoothes this area out. Take your foot off the foot pedal. Keep the nozzle over the end of the weld until the gas stops flowing. Keep the end of the filler metal in the shielding gas area.

12. Inspect the weld bead. Next, lay four weld beads on each side of two of the pieces of metal you have. Your second piece should have four good-quality beads on each side.

Name _____

13. On the next two pieces of metal, you will make weave beads. Start the welding arc in the same way as you did for laying the stringer bead. As the weld pool develops, move the torch from side to side to create a weld pool that is 3/8"–1/2" (9.5 mm–13 mm) wide.
    • For a weave bead, you will not add filler metal to the leading edge of the the weld pool, but to one side. Add filler metal to the side of the weld pool opposite the electrode. Refer to the Welding a Bead on Plate with Filler Metal section in Chapter 21 of the text.
    • Watch the width of the weld bead as you make the weld. Keep the weld bead the same width as you progress along the joint. Also, keep the amount of reinforcement, or buildup, the same.

14. Fill the weld pool at the end of the joint. Refer to step 11.

15. Inspect the weld beads. Next, weld four weave beads on each side of two of the pieces of metal you have. Your second piece should have four good-quality weave beads on each side.

16. To this point, you have learned to weld with a fairly constant current. Only slight changes in current could be made while welding. Now, set the current switch on the welding machine to the remote position. With the current switch in the remote position, the position of the foot pedal or thumb switch determines the amount of current used for welding. You can now change the amount of current considerably as you weld. This is very useful in controlling the weld pool and when ending a weld and filling the weld pool. As you approach the end of a weld bead, begin to decrease the current. Fill the weld pool with filler metal. Then, go over the area with the torch as before.

17. Repeat steps 7 through 14 on two pieces of the remaining metal. Inspect the completed weld beads. Repeat steps 7 through 14 for the remaining two pieces of metal.

18. You will show your instructor four pieces of the metal on which you have made welds. The first piece should have four stringer beads made with the current switch in the panel position. The second piece should have four weave beads, also made with the current switch in the panel position. The third piece should have four stringer beads made with the current switch in the remote position. The fourth piece should have four weave beads made with the current switch in the remote position.

## Inspection:

Each weld bead should be straight. The ripples should be slightly convex and evenly spaced. The width and reinforcement should remain the same along the weld bead. There should be no evidence of excessive penetration, porosity, or dipping of the electrode into the weld pool.

## Assigned Job 21-7
# Making Stringer and Weave Beads on Stainless Steel with Filler Metal

## Objectives:

In this job, you will learn to gas tungsten arc weld on stainless steel. You will strike a welding arc and move a molten weld pool along a plate. You will learn to control the weld pool, properly stop the welding arc, and fill the weld pool.

> **Note**
> Do not attempt this job until you have read all safety precautions, satisfactorily completed the *Gas Tungsten Arc Welding (GTAW) Safety Test*, and been approved by your instructor.

1. Obtain eight pieces of stainless steel that measure 1/8″ × 4″ × 6″ (3.2 mm × 100 mm × 150 mm).

2. Answer the following questions about welding 1/8″ (3.2 mm) stainless steel. Refer to the Selecting the Weld Current and Selecting and Preparing the Electrode sections and Figures 20-10, 20-13, and 20-19 in Chapter 20 of the text.

   A. What type of current is used? _____

   B. What type of electrode is used? _____

   C. What diameter of electrode is used? _____

   D. What diameter of filler metal is used? _____

   E. What type of filler metal is recommended? _____

   F. What amperage is used? (Use the amperage recommended for a butt weld.) _____

   G. What type of shielding gas is used? _____

   H. What is the recommended shielding gas flow rate? _____

3. Set up the welding machine using the following checklist. Refer to the Welding Machine Settings section in Chapter 20 of the text.
   - Set the type of electrical current: AC, DCEN (DCSP), or DCEP (DCRP). (See your answer to question 2A.)
   - Set the current range.
   - Set the desired current using the current control knob. (See your answer to question 2F.)
   - Set the high-frequency control for start only.
   - Set the contactor switch to the remote position.
   - Set the current switch to the panel position.
   - Set the post flow for about 10 seconds.

4. Obtain the correct electrode and 12 pieces of the correct filler metal. Prepare the electrode and install it in the torch. (See your answers to questions 2B, 2C, and 2D.)

5. Adjust the shielding gas flow for the proper rate. (See your answer to question 2H.)

6. Clean the metal to be welded. Use an awl or other sharp object and lightly scratch four straight lines into the metal. The lines should be about 1″ (25.4 mm) apart and 1/2″ (12.7 mm) from the edge. The lines should run the 6″ (150 mm) length of the metal. (See the drawing in step 7 of Assigned Job 21-6.) Place the metal in front of you on the work table.

7. Strike an arc. After the welding arc starts, move the electrode so it is 1/16″–1/8″ (1.5 mm–3.0 mm) from the work.

8. Hold the electrode in the same place until a molten weld pool forms. The weld pool should be about 3/16″–1/4″ (4.5 mm–6.0 mm) wide. While the weld pool is forming, bring the filler metal close to it.

9. As soon as the weld pool reaches the right size, add the filler metal to its front edge. Refer to the Welding a Bead on Plate with Filler Metal section in Chapter 21 of the text. After adding filler metal, begin moving forward. Keep the filler metal close to the weld pool. Watch the weld pool to keep it the same width as you move forward. Add filler metal as required to make a slightly convex weld bead.
   • If your electrode becomes contaminated, stop welding. Grind the electrode, reinstall it in the torch, and then continue welding.

10. When you reach the end of a weld, fill the weld pool with filler metal. Move the torch to the rear of the weld pool, and add the filler metal so it is the same height as the rest of the bead. Move the torch over this area, then take your foot off the foot pedal. If the current switch is in the remote position, gradually reduce the current. Refer to the Stopping the Arc and Welding a Bead on Plate with Filler Metal sections in Chapter 21 of the text. Keep the nozzle over the end of the weld until the gas stops flowing. Keep the end of the filler metal in the shielding gas area, as well.

11. Inspect the weld bead. Lay four weld beads on each side of two of the pieces of metal you have. Your second piece should have four good-quality weld beads on each side.

12. On the next two pieces of metal, you will make weave beads. Start the welding arc in the same way you did to lay the stringer beads. As the weld pool develops, move the torch from side to side to create a pool that is 3/8″–1/2″ (9.5 mm–13 mm) wide.
   • For a weave bead, you will not add the filler metal to the leading edge of the weld pool, but to one side. Add filler metal to the side of the pool opposite the electrode. Refer to the Welding a Bead on Plate with Filler Metal section in Chapter 21 of the text.
   • Keep the weld bead the same width as you progress. Also, keep the amount of reinforcement, or buildup, the same.

13. Fill the weld pool at the end of the weld. Refer to step 11.

14. Inspect the weld beads. Next, lay four weave beads on each side of two of the pieces of metal you have. Your second piece should have four high-quality weave beads on each side.

15. Repeat steps 7 through 10 using the remaining four pieces of metal. This time, place the current switch on the welding machine in the remote position.

16. You will show your instructor four pieces of metal. The first two pieces should be made with the current switch in the panel position. One should have four stringer beads; the second should have four weave beads. The third and fourth pieces should be made with the current switch in the remote position. One should have four stringer beads; the second should have four weave beads.

## Inspection:

Each weld bead should be straight, slightly convex, and have evenly spaced ripples. The width and reinforcement should remain the same along the weld bead. There should be no evidence of dipping the tungsten electrode into the weld pool or of porosity due to a very long welding arc length.

## Assigned Job 21-8
# Making Stringer and Weave Beads on Aluminum with Filler Metal

### Objectives:

In this job, you will learn to gas tungsten arc weld with high-frequency AC and filler metal. You will strike a welding arc and move a molten weld pool along a plate. In this job, you will use the foot pedal or thumb switch to control the amount of current. You will learn to control the weld pool, properly stop the welding arc, and fill the weld pool.

**Note**

Do not attempt this job until you have read all safety precautions, satisfactorily completed the *Gas Tungsten Arc Welding (GTAW) Safety Test*, and been approved by your instructor.

1. Obtain four pieces of aluminum that measure 1/16″ × 4″ × 6″ (1.6 mm × 100 mm × 150 mm).

2. Answer the following questions about welding 1/16″ (1.6 mm) aluminum. Refer to the Selecting the Weld Current, Selecting and Preparing the Electrode, and Filler Metals sections and Figures 20-9, 20-13, and 20-19 in Chapter 20.

   A. What type of current is used? _____

   B. What type of electrode is used? _____

   C. What diameter of electrode is used? _____

   D. What shape is the tip of an electrode used for aluminum welding? _____

   E. What diameter of filler metal is used? _____

   F. What type of filler metal is used? _____

   G. What amperage is used? (Treat a weld bead on plate as a butt joint to determine the amperage.) _____

   H. What type of shielding gas is used? _____

   I. What is the recommended shielding gas flow rate? _____

3. Set up the welding machine using the following checklist. Refer to the Welding Machine Settings section in Chapter 20 of the text.
   - Set the type of electrical current: AC, DCEN (DCSP), or DCEP (DCRP). (See your answer to question 2A.)
   - Set the current range.
   - Set the maximum desired current using the current control knob.
   - Set the high-frequency control for continuous.
   - Set the contactor switch to the remote position.
   - Set the current switch to the remote position.
   - Set the post flow for about 10 seconds.

4. Obtain the correct electrode, prepare the electrode, and install it into the torch. (See your answers to questions 2B, 2C, and 2D.)

5. Obtain eight pieces of the correct filler metal type and diameter. (See your answers to questions 2E and 2F.)

6. Adjust the shielding gas flow for the proper rate. (See your answer to question 2I)

7. Clean the metal to be welded by wire brushing to remove all surface oxides. Lightly scratch four straight lines into the metal. The lines should be about 1″ (25 mm) apart and 1/2″ (13 mm) from the edge. The lines should run the 6″ (152 mm) length of the metal. (See the drawing in step 6 of Assigned Job 21-1.) Place the metal in front of you on the work table.

8. Strike an arc. After the welding arc starts, move the electrode so it is 1/16″–1/8″ (1.5 mm–3.0 mm) from the work. With the current switch in the remote position, you can vary the current by adjusting the position of the foot pedal. For more current, press down on the foot pedal. For less current, let up on the foot pedal. You can still vary the current slightly by moving the electrode toward or away from the work.

9. Hold the electrode in the same place until a molten weld pool forms. The pool should be about 1/8″–3/16″ (3.0 mm–5.0 mm) wide. Remember, aluminum is one of the metals that has hot shortness. Do not let the weld pool become so large that it falls through and leaves a large hole.

10. While the weld pool is forming, bring the filler metal close to it.

11. When the pool becomes the right size, add the filler metal to its front edge. After adding the filler metal, begin moving the pool forward. Maintain the same size and convex shape for the length of the weld bead. If the weld pool gets too large, let up slightly on the foot pedal. To get more current for a larger weld pool, press down on the pedal.
    - Do not contaminate the electrode by dipping it into the weld or touching it to the filler metal. If you do, stop welding and grind off any contamination. Reinstall the electrode, form a ball on the end, and start welding again.

12. When you reach the end of a weld, fill the weld pool with filler rod. Move the torch to the rear of the weld pool and add filler metal. Move the torch over this area. Slowly let up on the foot pedal. This gradually reduces the current. Now, take your foot off the foot pedal. Keep the nozzle over the end of the weld and keep the filler metal in the shielding gas area until the gas stops flowing.

13. Inspect the weld bead. Lay four weld beads on each side of two of the pieces of metal. Your last piece should have four good-quality weld beads on each side.

14. Weld four weave beads on each side of the remaining two pieces of aluminum. Use the same techniques as for welding mild steel or stainless steel. Keep the end of the filler metal inside the shielding gas area or it will become oxidized. Fill the weld pool at the end of the weld as described in step 12.

## Inspection:

Each weld bead should be straight with ripples that are slightly convex and evenly spaced. The width should remain the same along the weld bead; the reinforcement should remain the same, as well. There should be no evidence of dipping the tungsten electrode into the weld pool or of porosity. Do not let the weld pool fall through the aluminum.

## Assigned Job 21-9
# Welding a Lap Joint with Filler Metal in the Flat Welding Position

### Objective:

In this job, you will learn to use filler metal to weld a lap joint in the flat welding position. The base metal that you will be welding will be selected by your instructor.

> **Note**
> Do not attempt this job until you have read all safety precautions, satisfactorily completed the *Gas Tungsten Arc Welding (GTAW) Safety Test*, and been approved by your instructor.

1. Obtain four pieces of the correct base metal for this job. Each piece should measure 1/8″ × 1 1/2″ × 6″ (3.2 mm × 40 mm × 150 mm).

2. Answer the following questions as they relate to the type of metal and metal thickness you will be welding. Refer to Chapter 20, especially Figures 20-8 through 20-19.

    A. What type of metal will you be welding? _____

    B. What type of current is used? _____

    C. What type of electrode is used? _____

    D. What diameter of electrode is used? _____

    E. What shape is the end of the electrode? _____

    F. What diameter of filler metal is used? _____

    G. What alloy filler metal is recommended? _____

    H. What amperage is used? _____

    I. What type of shielding gas is used? _____

    J. What is the recommended shielding gas flow rate? _____

3. Set up the welding machine using the following checklist. Refer to the Welding Machine Settings section in Chapter 20.
    - Set the type of electrical current: AC, DCEN (DCSP), or DCEP (DCRP). (See your answer to question 2B.)
    - Set the current range.
    - Set the desired current using the current control knob.
    - Set the high-frequency control to the start only position or the continuous position.
    - Set the contactor switch to the remote position.
    - Set the current switch to the desired position.
    - Set the post flow for about 10 seconds.

4. Answer the following questions about welding in the flat welding position. Refer to Figures 21-11 and 21-15 for proper angles.
   A. What is the work angle? _____
   B. What is the travel angle when forehand welding? _____
   C. What is the angle of the filler metal to the base metal? _____
   D. Where is the welding arc pointed? _____
   E. What is the shape of the weld pool? _____

5. Obtain the correct electrode, prepare the electrode, and install it in the torch. (See your answers to questions 2C, 2D, and 2E.)

6. Obtain two pieces of filler metal of the correct type and diameter. (See your answers to questions 2F and 2G.)

7. Adjust the shielding gas flow for the proper rate. (See your answer to 2J.)

8. Clean the metal to be welded with a wire brush. Place one piece of metal on another to form a lap joint, as shown in the following drawing. Tack weld the pieces in three places on each side of the joint. Place the tack-welded pieces in a weld positioner or support them so the joint is in the flat welding position.

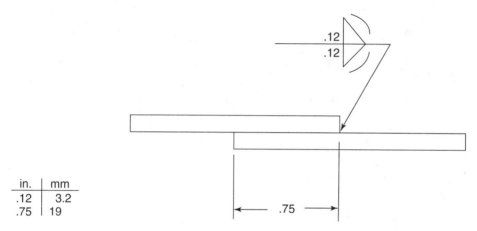

| in. | mm |
|-----|-----|
| .12 | 3.2 |
| .75 | 19 |

9. Strike an arc and develop a weld pool. Add filler metal as required to obtain a fillet weld that is convex and has a leg size equal to that shown on the AWS welding symbol. Weld a fillet weld and properly fill the weld pool with filler metal.

10. Inspect the fillet weld. Make any required changes on the welding machine. Weld a fillet weld on the opposite side of the metal. Position the metal so the joint is welded in the flat welding position.

11. Tack weld the other pieces of metal together and weld them as shown in the drawing. Weld all joints in the flat welding position.

12. Inspect the weld beads as stated in the following section. Your final piece should have good-quality fillet weld beads on each side.

## Inspection: _____

   Each fillet weld should be convex, with evenly spaced ripples. The size of the fillet should be as specified by the drawing. The weld bead should be the same height all along the joint. There should be no evidence of overlap, porosity, or dipping the electrode into the weld. Compare your completed weld to the one shown in Figure 21-14 in the text.

Name: _____  Date: _____

Class: _____  Instructor: _____

Lesson Grade: _____  Instructor's Initials: _____

## Assigned Job 21-10
# Welding a T-Joint with Filler Metal in the Flat Welding Position

## Objective

In this job, you will learn to weld a T-joint in the flat welding position using filler metal. The base metal that you will be welding will be selected by your instructor.

> **Note**
> Do not attempt this job until you have read all safety precautions, satisfactorily completed the *Gas Tungsten Arc Welding (GTAW) Safety Test*, and been approved by your instructor.

1. Obtain six pieces of the correct base metal for this job. Each piece should measure 1/8″ × 1 1/2″ × 6″ (3.2 mm × 40 mm × 150 mm). Obtain three additional pieces that measure 1/8″ × 3″ × 6″ (3.2 mm × 80 mm × 150 mm).

2. Answer the following questions as they relate to the type of metal and metal thickness you will be welding. Refer to Chapter 20, especially Figures 20-8 through 20-19.
   A. What type of metal will you be welding? _____
   B. What type of current is used? _____
   C. What type of electrode is used? _____
   D. What diameter of electrode is used? _____
   E. What shape is the end of the electrode? _____
   F. What diameter of filler metal is used? _____
   G. What alloy filler metal is recommended? _____
   H. What amperage is used? _____
   I. What type of shielding gas is used? _____
   J. What is the recommended shielding gas flow rate? _____

3. Set up the welding machine using the following checklist. Refer to the Welding Machine Settings section in Chapter 20 of the text.
   - Set the type of electrical current: AC, DCEN (DCSP), or DCEP (DCRP). (See your answer to question 2B.)
   - Set the current range.
   - Set the desired current.
   - Set the high-frequency control to the desired setting.
   - Set the contactor switch to the remote position.
   - Set the current switch to the desired position.
   - Set the post flow for about 10 seconds. A rule of thumb is 1 second of post flow for every 10 amps of welding current.

4. Answer the following questions about welding in the flat welding position. Refer to
   Figure 21-15 in the text for proper torch and filler metal angles.
   A. What is the proper work angle?_____
   B. What is the proper travel angle for forehand welding?_____
   C. What is the angle of the filler metal to the base metal? _____
   D. Where is the welding arc pointed? _____
   E. What is the shape of the weld pool? _____

5. Obtain the correct electrode, prepare the electrode, and install it in the torch. (See your answers
   to questions 2C, 2D, and 2E.)

6. Obtain two pieces of the correct filler metal. (See your answers to questions 2F and 2G.)

7. Adjust the shielding gas flow for the proper rate. (See your answer to question 2J.)

8. Clean the metal to be welded with a wire brush. Place one piece of metal on another to form a
   T-joint, as shown in Drawing A. Tack weld the pieces in three places on each side of the joint.
   Place the tack-welded pieces in a weld positioner or support the pieces so the joint is in the flat
   welding position.

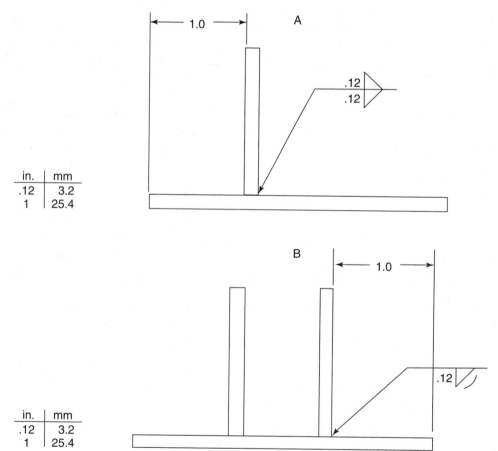

9. Strike the welding arc and develop a weld pool. Add filler metal as required to obtain a fillet
   weld that is convex and has the leg size shown in the AWS welding symbol. Weld a fillet weld
   and properly fill the weld pool with filler metal.

10. Inspect the fillet weld. Weld a fillet weld on the opposite side of the metal. Position the metal so
    the weld is made in the flat welding position.

Name _____

11. Tack weld a second smaller piece onto the assembly as shown in Drawing B. Weld a fillet weld on this joint.

12. Tack weld the other pieces of metal together and weld them as shown in the drawing. Weld all joints in the flat welding position.

13. Inspect the weld beads. Show your final welded assembly to your instructor for a grade.

## Inspection:

Each fillet weld should be convex, with ripples that are evenly spaced. The reinforcement and width of the weld bead should be the same along the length of the weld joint. There should be no evidence of any undercutting, overlap, or other defects in the weld beads.

## Assigned Job 21-11
# Welding a Square-Groove Butt Joint in the Flat Welding Position

## Objective:

In this job, you will learn to use GTAW and filler metal to weld a square-groove butt joint in the flat welding position. Your instructor will select the type of base metal you will weld.

> **Note**
> Do not attempt this job until you have read all safety precautions, satisfactorily completed the *Gas Tungsten Arc Welding (GTAW) Safety Test*, and been approved by your instructor.

1. Obtain eight pieces of the correct base metal for this job. Each piece should measure 1/8″ × 1 1/2″ × 6″ (3.2 mm × 40 mm × 150 mm).

2. Answer the following questions as they relate to the type and thickness of base metal you will be welding. Refer to Chapter 20, especially Figures 20-8 through 20-16 and Figure 20-19.

    A. What type of metal will you be welding? _____

    B. What type of current is used? _____

    C. What type of electrode is used? _____

    D. What diameter of electrode is used? _____

    E. What shape is the end of the electrode? _____

    F. What diameter of filler metal is used? _____

    G. What filler metal alloy is recommended? _____

    H. What amperage is used? _____

    I. What type of shielding gas is used? _____

    J. What is the recommended shielding gas flow rate? _____

3. Set up the welding machine using the following checklist. Refer to the Welding Machine Settings in Chapter 20.
    * Set the type of electrical current: AC, DCEN (DCSP), or DCEP (DCRP). (See your answer to question 2B.)
    * Set the current range.
    * Set the desired current.
    * Set the high-frequency control to the desired setting.
    * Set the contactor switch to the remote position.
    * Set the current switch to the desired position.
    * Set the post flow for about 10 seconds. A rule of thumb is 1 second of post flow for every 10 amps of welding current.

4. Answer the following questions about welding in the flat position.

   A. What is the proper work angle? _____

   B. What is the proper travel angle for forehand welding? _____

   C. What is the angle of the filler metal to the base metal? _____

   D. Where is the welding arc pointed? _____

   E. What is the shape of the opening at the root of the weld pool? _____

5. Obtain the correct electrode, prepare the electrode, and install it in the torch. (See your answers to questions 2C, 2D, and 2E.)

6. Obtain three pieces of the correct filler metal. (See your answers to questions 2F and 2G.)

7. Adjust the shielding gas flow for the proper rate. (See your answer to question 2J.)

8. Clean the base metal. Align two pieces of metal to form a butt weld as shown in the following drawing.

1/16″ = 1.6 mm

---

**Note**

When welding using direct current, use a root opening of 1/16″–3/32″ (1.5 mm–2.5 mm). When using AC, the root opening will be 3/32″–1/8″ (2.5 mm–3.0 mm).

---

9. Tack weld the pieces in three places. Check that the proper root opening still exists after tacking. Place the tack-welded pieces in a weld positioner or support the pieces so the joint is in the flat welding position.

10. Strike a welding arc and develop a weld pool. Watch the weld pool. Each piece of metal will begin to melt and the root opening will get slightly larger. This is the keyhole shape. It indicates that both pieces of metal are melting and you are obtaining full penetration. As the weld pool is developing, move the filler metal very close to it.

11. When the keyhole has formed, add the filler metal to the front edge of the weld pool. Begin moving the weld pool forward. As you move the weld pool, keep the keyhole shape the same size. Continue to add filler metal as required to fill the joint and to create a slight reinforcement on the surface. Inspect the completed weld.

12. Tack weld additional pieces of metal together and then weld them. Show your final three weldments to your instructor for a grade.

## Inspection:

A completed butt weld should have ripples that are evenly spaced, with the same amount of reinforcement along the weld. The weld bead should remain the same width along the weld joint. The weld root or back side of the weld should show complete penetration. There does not have to be a lot of metal on the back side of the joint. There should be no signs of porosity, undercutting, or inclusions along the joint.

Name: _____     Date: _____

Class: _____     Instructor: _____

Lesson Grade: _____     Instructor's Initials: _____

## Lesson 22
# GTAW: Horizontal, Vertical, and Overhead Welding Positions

## Objectives:

You will be able to identify the correct torch and filler metal angles for welding in the horizontal, vertical, and overhead welding positions. You will also be able to describe how welding out of position differs from welding in the flat welding position. You will be able to explain how to obtain good-quality welds when welding out of position.

## Instructions:

Read Chapter 22 and study Figures 22-1 through 22-13. Then, answer or complete the following questions.

_____    1.  To help prevent undercutting on a horizontal fillet weld, the torch is pointed more toward the _____.
   A.  horizontal piece
   B.  vertical piece
   C.  horizontal piece for forehand welds and the vertical piece for backhand welds
   D.  None of the above.

2.  *True or False?* Many welder qualification tests are done on out-of-position joints.

3.  When a butt joint is being welded out of position, what does the root opening look like?

_____

_____    4.  _____ welding current is used to weld overhead than is used to weld the same thickness in the flat position.
   A.  More
   B.  Less

5.  What is the proper shape of the weld pool when making a fillet weld out of position?

_____

6. Label the angles shown for welding a horizontal butt joint.

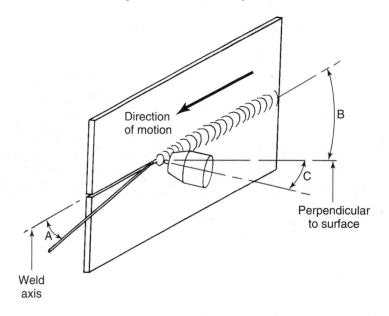

A. _____          C. _____

B. _____

7. *True or False?* The flow of argon gas must be decreased when welding in the overhead welding position because that gas will fall from the weld joint.

8. *True or False?* If the current set on the welding machine is reduced, the size of the weld pool and the amount of penetration will not change.

_____       9.   Thick metals welded in the vertical welding position are usually welded ____.
                  A.   uphill
                  B.   downhill

10. *True or False?* When making a horizontal fillet weld, porosity occurs when the weld pool sags down onto the horizontal piece.

Name: _____   Date: _____

Class: _____   Instructor: _____

Lesson Grade: _____   Instructor's Initials: _____

## Assigned Job 22-1
# Welding a Lap Joint in the Horizontal Welding Position

## Objective:

In this job, you will use the GTAW process to make a fillet weld on a lap joint in the horizontal welding position. The base metal that you will be welding will be selected by your instructor. Your instructor will also specify the size fillet to be welded.

**Note**

Do not attempt this job until you have read all safety precautions, satisfactorily completed the *Gas Tungsten Arc Welding (GTAW) Safety Test*, and been approved by your instructor.

1. Obtain six pieces of the correct base metal for this job. Four pieces should measure 1/8″ × 1 1/2″ × 6″ (3.2 mm × 40 mm × 150 mm), and two pieces should measure 1/8″ × 3″ × 6″ (3.2 mm × 80 mm × 150 mm). Obtain the correct electrode and two pieces of the correct filler metal.

2. Answer the following questions as they relate to the type and thickness of metal you will be welding. Refer to Chapter 20 of the text, especially Figures 20-8 through 20-16 and Figure 20-19.

   A. What type of metal will you be welding? _____

   B. What type of current is used? _____

   C. What type of electrode is used? _____

   D. What diameter of electrode is used? _____

   E. What shape is the end of the electrode? _____

   F. What diameter of filler metal is used? _____

   G. What alloy filler metal is recommended? _____

   H. What amperage is used? _____

   I. What type of shielding gas is used? _____

   J. What is the recommended shielding gas flow rate? _____

3. Use the following checklist to set up the welding machine, shielding gas, and torch. Refer to the Welding Machine Settings section in Chapter 20 of the text.
   - Verify that all connections are tight and that all cables are in good condition.
   - Set the type of electrical current: AC, DCEN (DCSP), or DCEP (DCRP).
   - Set the current range.
   - Set the desired current using the current control knob.
   - Set the high-frequency control for start only or continuous.
   - Set the contactor switch to the remote position.

- Set the current switch to the desired position.
- Set the post flow for about 10 seconds.
- Adjust the flow control to obtain the correct shielding gas flow rate.
- Prepare the electrode and install it in the torch.

4. Answer the following questions about welding in the horizontal position. For the proper angles, refer to Figure 22-3 in the text.
   A. What is the proper work angle? _____
   B. What is the proper travel angle? _____
   C. What is the angle of the filler metal to the base metal? _____
   D. Where is the torch pointed? _____
   E. What is the shape of the weld pool? _____

5. Clean the metal to be welded with a wire brush. Align the metal as shown in the following drawing. Tack weld the parts. Tack welding may be done in the flat welding position.

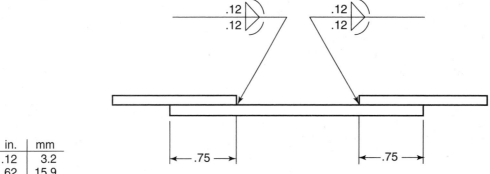

| in. | mm |
|-----|------|
| .12 | 3.2 |
| .62 | 15.9 |

6. Place the tack-welded parts in a weld positioner so the joint is in the horizontal welding position.

7. Weld the fillet welds indicated on the drawing, properly filling the weld pool at the end of each weld. Reposition the metal as needed so all welds are made in the horizontal welding position.

8. Tack weld the other pieces of metal together and weld them as shown. Weld each joint in the horizontal welding position. Show your final weldment to your instructor for a grade.

## Inspection:

Each fillet weld should be straight and slightly convex. The weld bead should have an even width and evenly spaced ripples. There should be no evidence of overlap, undercut, porosity, or dipping the electrode into the weld.

## Assigned Job 22-2
# Welding a T-Joint in the Horizontal Welding Position

## Objective:

In this job, you will use the GTAW process to make a fillet weld on a T-joint in the horizontal welding position. The base metal that you will be welding will be selected by your instructor.

> **Note**
> Do not attempt this job until you have read all safety precautions, satisfactorily completed the *Gas Tungsten Arc Welding (GTAW) Safety Test*, and been approved by your instructor.

1. Obtain four pieces of metal that measure 3/16″ × 1 1/2″ × 6″ (4.8 mm × 40 mm × 150 mm) and two pieces that measure 3/16″ × 3″ × 6″ (4.8 mm × 75 mm × 150 mm). Obtain the correct electrode and three pieces of the correct filler metal.

2. Answer the following questions as they relate to the type of metal and metal thickness you will be welding. Refer to Chapter 20 of the text, especially Figures 20-8 through 20-16 and Figure 20-19.

    A. What type of metal will you be welding? _____

    B. What type of current is used? _____

    C. What type of electrode is used? _____

    D. What diameter of electrode is used? _____

    E. What shape is the end of the electrode? _____

    F. What diameter of filler metal is used? _____

    G. What alloy filler metal is recommended? _____

    H. What amperage is used? _____

    I. What type of shielding gas is used? _____

    J. What is the recommended shielding gas flow rate? _____

3. Use the following checklist to set up the welding machine, shielding gas, and torch. Refer to the Welding Machine Settings section in Chapter 20 of the text.
    - Verify that all connections are tight and that all cables are in good condition.
    - Set the type of electrical current: AC, DCEN (DCSP), or DCEP (DCRP).
    - Set the current range.
    - Set the desired current using the current control knob.
    - Set the high-frequency control for start only or continuous.
    - Set the contactor switch to the remote position.
    - Set the current switch to the desired position.

- Set the post flow for about 10 seconds.
- Adjust the flow control to obtain the correct shielding gas flow rate.
- Prepare the electrode and install it in the torch.

4. Answer the following questions about welding in the horizontal welding position. Refer to Figure 22-3 in the text for proper angles.

   A. What is the proper work angle?_____

   B. What is the proper travel angle? _____

   C. What is the angle of the filler metal to the base metal? _____

   D. Where is the torch pointed? _____

   E. What is the shape of the weld pool? _____

5. Clean the metal to be welded with a wire brush. Place one of the smaller pieces onto a larger piece as shown in the following drawing. Tack weld this joint in three places on each side of the joint.

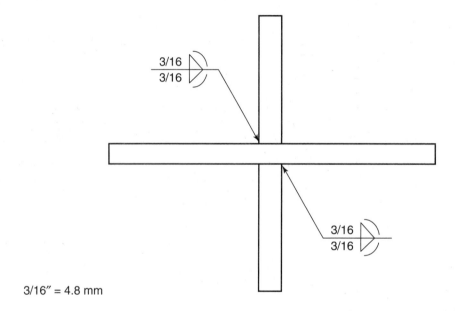

3/16" = 4.8 mm

6. Place the tack-welded metal so the joint is in the horizontal welding position.

7. Weld a fillet weld on both sides of the metal, as shown on the drawing. Both welds are to be done in the horizontal welding position. Refer to Figure 22-3 in the text for the proper torch and filler metal angles.

8. Tack weld a second smaller piece onto the assembly. Weld each joint in the horizontal welding position. Inspect these welds.

9. Repeat this welding exercise with the remaining material, welding all joints in the horizontal welding position. Show your final weldment to your instructor for a grade.

## Inspection:

Each fillet weld should be straight and slightly convex. The weld bead should have an even width and evenly spaced ripples. There should be no evidence of overlap, undercut, porosity, or dipping the electrode into the weld.

## Assigned Job 22-3
# Welding a Square-Groove Butt Joint in the Horizontal Welding Position

## Objective:

In this job, you will weld a square-groove butt joint in the horizontal welding position. Your instructor will select the base metal that you will be welding.

> **Note**
> Do not attempt this job until you have read all safety precautions, satisfactorily completed the *Gas Tungsten Arc Welding (GTAW) Safety Test*, and been approved by your instructor.

1. Obtain six pieces of the correct base metal for this job. Each piece should measure $1/8'' \times 1\,1/2'' \times 6''$ (3.2 mm × 40 mm × 150 mm). Obtain the correct electrode and two pieces of the correct filler metal.

2. Answer the following questions as they relate to the type of metal and metal thickness you will be welding. Refer to Chapter 20 of the text, especially Figures 20-8 through 20-16 and Figure 20-19.

   A. What type of metal will you be welding? _____

   B. What type of current is used? _____

   C. What type of electrode is used? _____

   D. What diameter of electrode is used? _____

   E. What shape is the end of the electrode? _____

   F. What diameter of filler metal is used? _____

   G. What alloy filler metal is recommended? _____

   H. What amperage is used? _____

   I. What type of shielding gas is used? _____

   J. What is the recommended shielding gas flow rate? _____

3. Use the following checklist to set up the welding machine, shielding gas, and torch. Refer to the Welding Machine Settings section in Chapter 20.
   - Verify that all connections are tight and that all cables are in good condition.
   - Set the type of electrical current, the current range, and the desired current.
   - Set the high-frequency control to the desired position.
   - Set the contactor switch to the remote position.
   - Set the current switch to the desired position.
   - Set the post flow and adjust the flow control for the correct shielding gas flow rate.
   - Prepare the electrode and install it in the torch.

4. Answer the following questions about welding in the horizontal welding position. For the proper angles, refer to Figure 22-6 in the text.

   A. What is the proper work angle?_____

   B. What is the proper travel angle? _____

   C. What is the angle of the filler metal to the base metal? _____

   D. Where is the welding arc pointed? _____

   E. What is the shape of the root opening? _____

5. Align two pieces of metal so there is a 1/16″ (1.5 mm) gap between them. Make three tack welds along the joint. After tacking, the gap should still measure 1/16″ (1.5 mm) along the entire length. Tack weld a third piece onto the first two as shown in the following drawing.

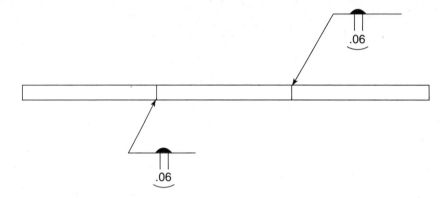

.06″ = 1.6 mm

6. Place the metal in a weld positioner so the joint is in the horizontal welding position. Weld the butt joint as shown in the drawing. Watch the weld pool as you weld. Do not allow the weld pool to sag. If it begins to sag, change the upward torch angle, reduce the current, or do both. To ensure 100% penetration, the gap at the root of the weld pool should look like a keyhole. The keyhole indicates that both base metal pieces are melting. Keep the keyhole the same size as you move the weld pool along the weld axis.

7. Tack weld the remaining pieces and then weld them in the horizontal position. After completing each weld, examine it and make any corrections necessary.

## Inspection:

Each butt weld should be straight and slightly convex. It should have ripples that are evenly spaced. There should not be any sag of the weld onto the lower piece. This causes overlap. There should be no areas of underfill, undercut, or other defects. The penetration on the back side of the weld should be flush to slightly convex.

Name: _____  Date: _____

Class: _____  Instructor: _____

Lesson Grade: _____  Instructor's Initials: _____

## Assigned Job 22-4
# Welding a Lap Joint in the Vertical Welding Position

## Objective:

In this job, you will use the GTAW process to make a fillet weld on a lap joint in the vertical welding position. The base metal that you will be welding will be selected by your instructor.

> **Note**
> Do not attempt this job until you have read all safety precautions, satisfactorily completed the *Gas Tungsten Arc Welding (GTAW) Safety Test*, and been approved by your instructor.

1. Obtain six pieces of the correct base metal for this job. Each piece should measure 1/8″ × 1 1/2″ × 6″ (3.2 mm × 40 mm × 150 mm). Obtain the correct electrode and three pieces of the correct filler metal.

2. Answer the following questions as they relate to the type of metal and metal thickness you will be welding. Refer to Chapter 20 of the text, especially Figures 20-8 through 20-16 and Figure 20-19.

   A. What type of metal will you be welding? _____

   B. What type of current is used? _____

   C. What type of electrode is used? _____

   D. What diameter of electrode is used? _____

   E. What shape is the end of the electrode? _____

   F. What diameter of filler metal is used? _____

   G. What alloy filler metal is recommended? _____

   H. What amperage is used? _____

   I. What type of shielding gas is used? _____

   J. What is the recommended shielding gas flow rate? _____

3. Use the following checklist to set up the welding machine, shielding gas, and torch. Refer to the Welding Machine Settings section in Chapter 20 of the text.
   - Verify that all connections are tight and that all cables are in good condition.
   - Set the type of electrical current, the current range, and the desired current.
   - Set the high-frequency control to the desired position.
   - Set the contactor switch to the remote position.
   - Set the current switch to the desired position.
   - Set the post flow and adjust the flow control for the correct shielding gas flow rate.
   - Prepare the electrode and install it in the torch.

4. Answer the following questions about welding in the vertical position. Refer to Figure 22-9 in the text for the proper angles.

   A. What is the proper work angle? _____

   B. What is the proper travel angle? _____

   C. What is the angle of the filler metal to the base metal? _____

   D. Where is the torch pointed? _____

   E. What is the shape of the weld pool? _____

5. Clean the metal to be welded with a wire brush. Arrange the metal as shown in the following drawing. Tack weld the parts. Tack welding may be done in the flat welding position.

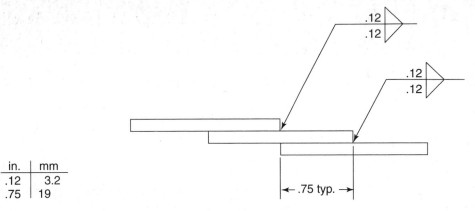

| in. | mm |
|-----|-----|
| .12 | 3.2 |
| .75 | 19 |

6. Place the tack-welded parts in a weld positioner so the joint to be welded is in the vertical welding position.

7. Weld the fillet welds as indicated on the drawing, properly filling the weld crater at the end of each weld. Reposition the metal as required so all welds are made in the vertical position. Refer to Figure 22-9 in the text for the proper torch angles. Weld in the vertically down direction.

8. Tack weld the other pieces of metal together and weld them as shown in the drawing. Weld all joints in the vertical position. Show your final weldment to your instructor for a grade.

## Inspection:

Each fillet weld should be straight and slightly convex. The weld bead should have an even width and evenly spaced ripples. There should be no evidence of undercut, porosity, or dipping the electrode into the weld.

## Assigned Job 22-5
# Welding a T-Joint in the Vertical Welding Position

## Objective:

In this job, you will use the GTAW process to make a fillet weld on a T-joint in the vertical welding position. The base metal that you will be welding will be selected by your instructor.

> **Note**
>
> Do not attempt this job until you have read all safety precautions, satisfactorily completed the *Gas Tungsten Arc Welding (GTAW) Safety Test*, and been approved by your instructor.

1.  Obtain four pieces of the selected base metal that measure 1/8″ × 1 1/2″ × 6″ (3.2 mm × 40 mm × 150 mm) and four pieces that measure 3/16″ × 1 1/2″ × 6″ (4.8 mm × 40 mm × 150 mm). Obtain the correct electrode and four pieces of the correct filler metal.

2.  Answer the following questions as they relate to the type of metal and metal thickness you will be welding. Refer to Chapter 20 in the text, especially Figures 20-8 through 20-16 and Figure 20-19.

    A.  What type of metal will you be welding? _____

    B.  What type of current is used? _____

    C.  What type of electrode is used? _____

    D.  What diameter of electrode is used? _____

    E.  What shape is the end of the electrode? _____

    F.  What diameter of filler metal is used? _____

    G.  What alloy filler metal is recommended? _____

    H.  What amperage is used? _____

    I.  What type of shielding gas is used? _____

    J.  What is the recommended shielding gas flow rate? _____

3.  Use the following checklist to set up the welding machine, shielding gas, and torch. Refer to the Welding Machine Settings section in Chapter 20 of the text.
    -   Verify that all connections are tight and that all cables are in good condition.
    -   Set the type of electrical current, the current range, and the desired current.
    -   Set the high-frequency control to the desired position.
    -   Set the contactor switch to the remote position.
    -   Set the current switch to the desired position.
    -   Set the post flow and adjust the flow control for the correct shielding gas flow rate.
    -   Prepare the electrode and install it in the torch.

4. Answer the following questions about welding in the vertical position. Refer to Figure 22-9 in the text for proper angles.

   A. What is the proper work angle? _____

   B. What is the proper travel angle? _____

   C. What is the angle of the filler metal to the base metal? _____

   D. Where is the torch pointed? _____

   E. What is the shape of the weld pool? _____

5. Place one of the 1/8″ (3.2 mm) pieces on the other 1/8″ (3.2 mm) piece to form the weldment shown in the following drawing. Tack weld in three places on each side of the joint.

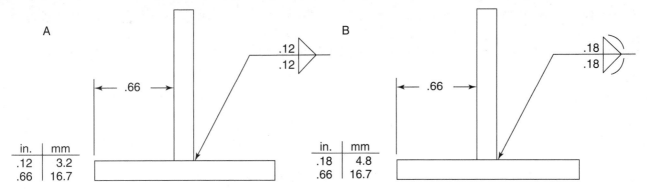

6. Place the tack-welded metal in a positioner so the joint is in the vertical welding position.

7. Weld the fillet welds shown in part A of the drawing. Both welds are to be made in the vertical welding position.

8. Repeat steps 5 and 6 using the 3/16″ (4.8 mm) pieces.

9. Weld a fillet weld on both sides of the 3/16″ (4.8 mm) metal as shown in B. This weld may require more than one pass to complete. Each weld bead must melt into the base metal and the previous weld bead. You can use stringer beads or a weaving motion, but keep the weld pool small so it does not sag.

## Inspection:

Each fillet weld should be straight and slightly convex. The weld beads should be even in width and have evenly spaced ripples. Fillet leg sizes should be equal and of the sizes specified in the drawing. There should be no evidence of undercut, porosity, or dipping the electrode into the weld. A weld made in the vertical welding position should look as good as a weld done in the flat welding position.

## Assigned Job 22-6
# Welding a Square-Groove Butt Joint in the Vertical Welding Position

## Objective:

In this job, you will weld a square-groove butt joint in the vertical welding position. Your instructor will select the base metal that you will weld.

> **Note**
> Do not attempt this job until you have read all safety precautions, satisfactorily completed the *Gas Tungsten Arc Welding (GTAW) Safety Test*, and been approved by your instructor.

1. Obtain six pieces of the correct base metal for this job. Each piece should measure 1/8″ × 1 1/2″ × 6″ (3.2 mm × 40 mm × 150 mm). Obtain the correct electrode and two pieces of the correct filler metal.

2. Answer the following questions as they relate to the type of metal and metal thickness you will be welding. Refer to Chapter 20 in the text, especially Figures 20-8 through 20-16 and Figure 20-19.

   A. What type of metal will you be welding? _____

   B. What type of current is used? _____

   C. What type of electrode is used? _____

   D. What diameter of electrode is used? _____

   E. What shape is the end of the electrode? _____

   F. What diameter of filler metal is used? _____

   G. What alloy filler metal is recommended? _____

   H. What amperage is used? _____

   I. What type of shielding gas is used? _____

   J. What is the recommended shielding gas flow rate? _____

3. Use the following checklist to set up the welding machine, shielding gas, and torch. Refer to the Welding Machine Settings section in Chapter 20 of the text.
   - Verify that all connections are tight and that all cables are in good condition.
   - Set the type of electrical current, the current range, and the desired current.
   - Set the high-frequency control, contactor switch, and current switch to the desired positions.
   - Set the post flow and adjust the flow control for the correct shielding gas flow rate.
   - Prepare the electrode and install it in the torch.

4. Answer the following questions about welding in the vertical welding position.

   A. What is the proper work angle? _____

   B. What is the proper travel angle? _____

   C. What is the angle of the filler metal to the base metal? _____

   D. Where is the welding arc pointed? _____

   E. What is the shape of the root opening? _____

5. Align two pieces of metal so there is a 1/16″ (1.5 mm) gap between them. The root opening for AC welding may be slightly larger, about 3/32″ (2.5 mm). Make three tack welds along the joint. After tacking, the gap should still measure 1/16″ (1.5 mm) or 3/32″ (2.5 mm) along the entire length. Tack weld a third piece onto the first two as shown in the following drawing.

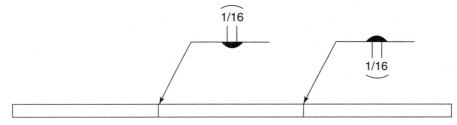

1/16″ = 1.6 mm

6. Place the metal in a weld positioner so the weld can be made in the vertical welding position. Weld the butt joints as shown in the drawing. Watch the weld pool as you weld. Do not allow the weld pool to sag. If it begins to sag, change the upward torch angle, reduce the current, or both. To ensure 100% penetration, the gap at the root of the weld pool should look like a keyhole. The keyhole indicates that both pieces of the base metal are melting. Keep this keyhole the same size as the weld progresses.

7. Tack weld the remaining pieces and then weld them. After completing each weld, examine it and make any corrections needed.

## Inspection:

Each butt weld should be straight, slightly convex, and have evenly spaced ripples. The penetration on the back side of the weld should be flush to slightly convex. It should be consistent along the joint. There should be no defects along the weld or excessive penetration on the back side of the weld.

# Assigned Job 22-7
# Welding a Lap Joint in the Overhead Welding Position

## Objective:

In this job, you will use the GTAW process to make a fillet weld on a lap joint in the overhead welding position. Your instructor will select the type of base metal you will weld.

> **Note**
>
> Do not attempt this job until you have read all safety precautions, satisfactorily completed the *Gas Tungsten Arc Welding (GTAW) Safety Test*, and been approved by your instructor.

1. Obtain six pieces of the correct base metal for this job. Each piece should measure 1/8″ × 1 1/2″ × 6″ (3.2 mm × 40 mm × 150 mm). Obtain the correct electrode and three pieces of the correct filler metal.

2. Answer the following questions as they relate to the type of metal and metal thickness you will be welding. Refer to Chapter 20 of the text, especially Figures 20-8 through 20-16 and Figure 20-19. Also refer to the Welding in the Overhead Welding Position section in Chapter 22.

   A. What type of metal will you be welding? _____

   B. What type of current is used? _____

   C. What type of electrode is used? _____

   D. What diameter of electrode is used? _____

   E. What shape is the end of the electrode? _____

   F. What diameter of filler metal is used? _____

   G. What alloy filler metal is recommended? _____

   H. What amperage is used? _____

   I. What type of shielding gas is used? _____

   J. What is the recommended shielding gas flow rate? _____

3. Use the following checklist to set up the welding machine, shielding gas, and torch. Verify that everything is operating properly. Refer to the Welding Machine Settings section in Chapter 20 of the text.
   - Set the type of electrical current, the current range, and the desired current.
   - Set the high-frequency control, contactor switch, and the current switch to the desired positions.
   - Set the post flow and adjust the flow control for the correct shielding gas flow rate.
   - Prepare the electrode and install it in the torch.

4. Answer the following questions about welding in the overhead position.

   A. What is the proper work angle? _____

   B. What is the proper travel angle? _____

   C. What is the angle of the filler metal to the base metal? _____

   D. Where is the torch pointed? _____

   E. What is the shape of the weld pool? _____

5. Align two pieces of metal (pieces A and B) as shown in the following drawing. Tack weld each lap joint in three places. This may be done in the flat welding position.

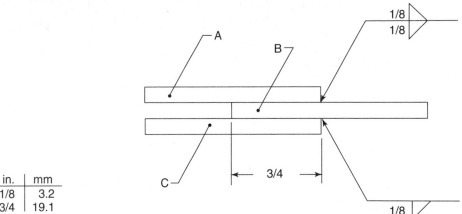

| in. | mm |
|-----|------|
| 1/8 | 3.2 |
| 3/4 | 19.1 |

6. Place the tack-welded parts in a positioner so the joint to be welded is in the overhead welding position.

7. Weld the fillet welds as indicated on the drawing. Each weld is to be done in the overhead welding position.

**Note**

Turn the metal as required so each weld is made in the overhead welding position. Keep the weld pool small. There can be no overlap of weld metal onto the surface of piece B, or piece C will not fit tightly.

8. Tack weld piece C into place. Weld a fillet weld in the overhead welding position.

9. Repeat this exercise with the remaining material for a grade.

## Inspection:

   Each fillet weld should be straight and slightly convex. The weld beads should have consistent widths and evenly spaced ripples. There should be no evidence of undercut, porosity, or dipping the electrode into the weld.

Name: _____ Date: _____

Class: _____ Instructor: _____

Lesson Grade: _____ Instructor's Initials: _____

## Assigned Job 22-8
# Welding an Inside Corner Joint in the Overhead Welding Position

## Objective:

In this job, you will use the GTAW process to make a fillet weld on an inside corner joint in the overhead welding position. The base metal that you will be welding will be selected by your instructor.

> **Note**
> Do not attempt this job until you have read all safety precautions, satisfactorily completed the *Gas Tungsten Arc Welding (GTAW) Safety Test*, and been approved by your instructor.

1. Obtain six pieces of the selected base metal that measure 1/8″ × 1 1/2″ × 6″ (3.2 mm × 40 mm × 150 mm) and three pieces that measure 1/8″ × 3″ × 6″ (3.2 mm × 75 mm × 150 mm). Obtain the correct electrode and four pieces of the correct filler metal.

2. Answer the following questions as they relate to the type of metal and metal thickness you will be welding. Refer to Chapter 20 in the text, especially Figures 20-8 through 20-16 and Figure 20-19.

   A. What type of metal will you be welding? _____

   B. What type of current is used? _____

   C. What type of electrode is used? _____

   D. What diameter of electrode is used? _____

   E. What shape is the end of the electrode? _____

   F. What diameter of filler metal is used? _____

   G. What alloy filler metal is recommended? _____

   H. What amperage is used? _____

   I. What type of shielding gas is used? _____

   J. What is the recommended shielding gas flow rate? _____

3. Use the following checklist to set up the welding machine, shielding gas, and torch. Verify that everything is operating properly. Refer to the Welding Machine Settings section in Chapter 20 of the text.
   - Set the type of electrical current, the current range, and the desired current.
   - Set the high-frequency control, contactor switch, and current switch to the desired positions.
   - Set the post flow and adjust the flow control for the correct shielding gas flow rate.
   - Prepare the electrode and install it in the torch.

4. Answer the following questions about welding a fillet weld in the overhead position.

   A. What is the proper work angle? _____

   B. What is the proper travel angle? _____

   C. What is the angle of the filler metal to the base metal? _____

   D. Where is the torch pointed? _____

   E. What is the shape of the weld pool? _____

5. Align and tack weld the pieces to form the weldment shown in the following drawing. Tack welding can be done in the flat welding position.

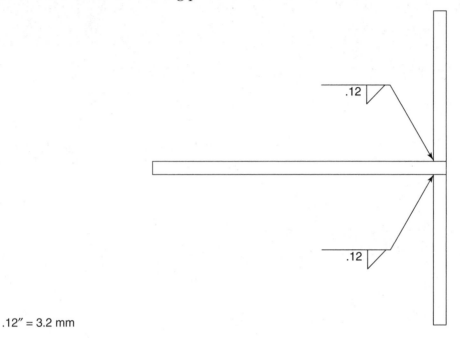

.12″ = 3.2 mm

6. Place the tack-welded metal in a weld positioner so the joint to be welded is in the overhead welding position.

7. Make the fillet welds as shown on the drawing. Both welds are to be done in the overhead welding position.

8. Repeat this exercise with the remaining material. Evaluate each fillet weld after completing it. Check for porosity. The Welding in the Overhead Welding Position section in Chapter 22 of the text discusses what to do if porosity is present. Make any necessary changes before making the next weld. Repeat this job with the remaining material. Show the final two weldments to your instructor for a grade.

## Inspection:

Fillet weld edges should be straight and even in width. The weld face should be slightly convex. There should be evenly spaced ripples in the weld bead. There should be no evidence of undercut, porosity, or dipping the electrode into the weld.

## Assigned Job 22-9
# Welding Square-Groove and V-Groove Butt Joints in the Overhead Welding Position

## Objective:

In this job, you will weld a square-groove butt joint and a V-groove butt joint on an outside corner joint in the overhead welding position. Your instructor will select the base metal that you will weld.

> **Note**
> Do not attempt this job until you have read all safety precautions, satisfactorily completed the *Gas Tungsten Arc Welding (GTAW) Safety Test*, and been approved by your instructor.

1. Obtain six pieces of correct base metal for this job. Each piece should measure 1/8″ × 1 1/2″ × 6″ (3.2 mm × 40 mm × 150 mm). Obtain the correct electrode and two pieces of the correct filler metal.

2. Answer the following questions as they relate to the type of metal and metal thickness you will be welding. Refer to Chapter 20 in the text, especially Figures 20-8 through 20-16 and Figure 20-19.

    A. What type of metal will you be welding? _____

    B. What type of current is used? _____

    C. What type of electrode is used? _____

    D. What diameter of electrode is used? _____

    E. What shape is the end of the electrode? _____

    F. What diameter of filler metal is used? _____

    G. What alloy filler metal is recommended? _____

    H. What amperage is used? _____

    I. What type of shielding gas is used? _____

    J. What is the recommended shielding gas flow rate? _____

3. Use the following checklist to set up the welding machine, shielding gas, and torch. Refer to the Welding Machine Settings section in Chapter 20 of the text.
    - Set the type of electrical current, the current range, and the desired current.
    - Set the high-frequency control, contactor switch, and current switch to the desired position.
    - Set the post flow and adjust the flow control for the correct shielding gas flow rate.
    - Prepare the electrode and install it in the torch.

4.  Answer the following questions about welding in the overhead welding position.

    A.  What is the proper work angle? _____

    B.  What is the proper travel angle? _____

    C.  What is the angle of the filler metal to the base metal? _____

    D.  Where is the arc pointed? _____

    E.  What is the shape of the root opening? _____

5.  Align two pieces of metal so there is a 1/16″ (1.5 mm) gap between them. The root opening for AC welding may be slightly larger, or about 3/32″ (2.5 mm). Make three tack welds along the joint. After tacking, the gap should still measure 1/16″ (1.5 mm) or 3/32″ (2.5 mm) along the entire length. Tack weld a third piece onto the first two as shown in the following drawing.

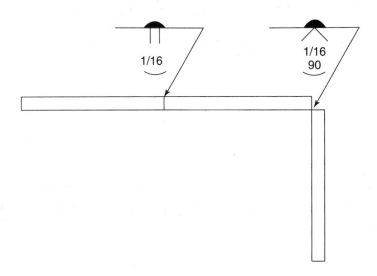

1/16″ = 1.6 mm

6.  Place the metal in a weld positioner so the butt weld can be made in the overhead welding position. Weld the butt joint as shown in the drawing. Rotate the metal so the V-groove weld can be made in the overhead welding position. Fill the V-groove on the outside corner joint from edge to edge.

**Note**
As you weld, the root of the weld pool should look like a keyhole. Keep this keyhole the same size as you make the weld.

7.  Repeat steps 5 and 6 for the remaining pieces and then weld them. After making each weld, examine it and make any necessary corrections.

## Inspection:

Each butt weld should be straight, slightly convex, and have evenly spaced ripples. The penetration must be 100%, flush to slightly convex, and consistent along the joint. There should be no defects along the weld or excessive penetration on the back side of the weld. The outside corner joint should be filled from edge to edge, have 100% penetration, and exhibit no defects.

# Section 5
# Plasma Arc Cutting

## Plasma Arc Cutting Safety Test

## Objectives:

You will be able to discuss the potential safety hazards of plasma arc cutting. You will be able to describe the safety precautions required when working with plasma arc cutting equipment.

## Instructions:

Always follow safe practices when cutting. If you have safety questions or concerns, ask your instructor. This test does not include questions about every safety topic, but is intended to highlight key items. Review Chapter 23 and Figures 23-1 through 23-15. Then, answer or complete the following questions.

1. *True or False?* The presence of an arc can cause safety hazards.

2. What part of the torch is used to start and stop the arc?

   _____

_____ 3. Temperatures used during the plasma arc cutting process range between _____.
   A   2500°F–3500°F (1400°C–1900°C)
   B.   5000°F–7500°F (2700°C–4100°C)
   C.   10,000°F–13,000°F (5500°C–7200°C)
   D.   18,000°F–25,000°F (10,000°C–14,000°C)

_____ 4. Which of the following is *not* one of the hazards present during PAC?
   A.   Intense light.
   B.   Intense sound.
   C.   High-pressure acetylene.
   D.   Flying particles.

5. List two methods to protect your hearing during plasma arc cutting.

   _____

   _____

6. Is ventilation required when PAC? If yes, why?

   _____

   _____

7. What types of protective clothing should be worn for PAC?

_____

_____ 8. What is the minimum filter lens to be used when PAC?
  A.  A #8 filter lens.
  B.  A #9 filter lens.
  C.  A #10 filter lens.
  D.  A #12 filter lens.

9. *True or False?* Plasma arc cutting is a combination of an arc process and a cutting process. The dangers of both are present.

10. *True or False?* A piercing operation produces less upward flying particles than a cut on thin steel.

# Lesson 23
# Plasma Arc Cutting

## Objective:
You will be able to describe the operation and setup of plasma arc welding equipment.

## Instructions:
Review Chapter 23 and Figures 23-1 through 23-15. Then, answer or complete the following questions.

1. *True or False?* Plasma is a state of matter.

2. What arc welding processes have plasma present?

   _____

3. In PAC, the transferred arc is created between the _____ and the _____.

   _____

   _____

_____  4. Which of the following is *not* an advantage of PAC?
   - A. The ability to cut all metals.
   - B. Electricity is required.
   - C. Distortion is minimized.
   - D. No explosive gases are used.

5. *True or False?* A constant voltage machine is used for PAC.

_____  6. Which of the following metals is *not* used to make PAC electrodes?
   - A. Thorium.
   - B. Halfnium.
   - C. Zirconium.
   - D. Tungsten.

7. The plasma gas is forced through a small hole in what part of the PAC torch?

   _____

8. Name the two categories of gas used in plasma arc cutting.

   _____

   _____

*For Questions 9–11, write the best gas combinations for each of the cutting applications.*

9. Excellent cut on 3/4" (19 mm) thick stainless steel.

_____

10. Good cut quality and speed on mild steel, stainless steel, and aluminum.

_____

11. Excellent cut quality and speed on mild steel.

_____

_____ 12. What gas is most commonly used for manual plasma arc cutting?
   A.   Air.
   B.   Argon.
   C.   Argon with 2% oxygen.
   D.   Helium with 35% nitrogen.

13. Label the parts of the PAC torch shown:

   A. _____

   B. _____

   C. _____

   D. _____

   E. _____

14. A minimum of a #_____ filter plate or lens should be used for PAC.

_____

_____ 15. What creates the loud noise sometimes present with PAC?
   A.   The inverter power supply.
   B.   Improperly assembled electrode and nozzle.
   C.   High-amperage cutting arc.
   D.   High-velocity gas.

16. What is the purpose of the pilot arc?

_____

17. The arc between the electrode and the base metal is called a(n) _____ arc.

_____

18. What two parts of the torch are consumable items?

_____

_____

19. What is the cause of dross that is hard to remove?

_____

20. A travel angle of _____° is used for gouging.

_____

Name: _____     Date: _____

Class: _____     Instructor: _____

Lesson Grade: _____     Instructor's Initials: _____

## Assigned Job 23-1
# Assembling Plasma Arc Cutting Equipment

## Objectives:

In this job, you will learn the parts of a plasma arc welding station. You will be able to assemble a plasma arc welding torch.

---

**Note**

Do not attempt this job until you have read all safety precautions, satisfactorily completed the *Plasma Arc Cutting (PAC) Safety Test*, and been approved by your instructor.

---

1. Your instructor will assign you plasma arc cutting equipment to use for this job. Examine the equipment, including the nameplate, and answer the following questions.

    A. What company manufactures the power supply? _____

    B. What voltage does the machine require? _____

    C. What is the duty cycle (if listed)? _____

    D. Is the machine hard wired into an electronic panel or box or is it plugged into an electrical outlet? _____

    E. Who is the manufacturer of the cutting torch? _____

2. Connect a source of shielding gas from a cylinder or manifold if one is not already connected. If shielding gas is connected, verify that the connections are tight. Do not overtighten the connections. If there is no place for shielding gas to be connected, verify with your instructor that the machine has an internal air compressor.

3. Connect a workpiece lead to the power supply. Inspect the workpiece lead. There should be no cuts or damage to the cable.

4. Connect the combination cable to the power supply. If the cable is already connected to the welding machine, remove and reattach the combination cable. This is usually done by plugging the cable in and securing it in place by turning a threaded collar.

5. Inspect the combination cable for any signs of cuts or wear.

6. Remove the heat shield, constricting nozzle, and electrode from the torch. Look at the electrode. Notice the color of the center of the electrode assembly.

   A. List the parts that you removed from the torch.

   _____

   _____

   _____

7. Reinstall the electrode, constricting nozzle, and heat shield into the torch. Depending on the manufacturer of your torch, there may be additional parts to be installed. Refer to step 6A when reassembling the torch.

8. You should now be familiar with the PAC equipment. Ask your instructor to look over the equipment and verify that everything is in good condition.

## Assigned Job 23-2
# Plasma Arc Cutting Mild Steel

## Objectives:

In this job, you will learn to use the plasma arc cutting equipment to make straight cuts on mild steel. Your instructor will assign you plasma arc cutting equipment to use for this job. Your instructor will also select the type of plasma gas, shielding gas, and electrode you will be using.

> **Note**
> Do not attempt this job until you have read all safety precautions, satisfactorily completed the *Plasma Arc Cutting (PAC) Safety Test*, and been approved by your instructor.

1. Obtain two pieces of mild steel 1/8″ × 6″ × 6″ (3.2 mm × 150 mm × 150 mm).

2. Mark lines every 3/8″ (9.5 mm). The lines can be drawn with soapstone or chalk or scratched with a sharp object.

3. Answer the following questions about cutting mild steel with the equipment you have been assigned.

   A. What plasma gas will be used? _____

   B. What shielding gas will be used? _____

   C. Refer to Figure 23-7 in the text. What results are expected: excellent, good, or fair? _____

   D. What type of electrode will be used? _____

4. Refer to Figure 23-10 in the text and answer the following questions about cutting 1/8″ (3.2 mm) mild steel.

   A. What current is recommended? _____

   B. What travel speed is recommended? _____

5. Inspect your work area and remove any flammable items.

6. Assemble the equipment and install the electrode into the torch.

7. Connect the workpiece lead to the work.

8. Turn on the PAC equipment. Open the gas cylinder valve or the valve to the shop manifold system if shielding gas is used. Adjust the regulator to the correct pressure, often about 65 psig. Turn on the water supply if used.

9. Set the correct current on the power supply. If you are not sure of the exact setting, set the current slightly high to ensure a complete cut.

10. Hold the torch 1/16"–1/8" (1.5 mm–3.0 mm) from the surface of the work. Hold the torch over the end of the metal.

11. Press the switch to start the cut. The arc will start and rapidly melt the base metal. Start moving forward. Plasma arc cutting is a fast cutting process on thin metals. You will need to move quickly. See the recommended travel speed recorded in step 4.

12. If the arc does not establish, release the switch. Make sure you are holding the torch over the edge of the work and the correct distance from the work. Pull the trigger or switch and start the cut.

13. Move along the line on the base metal. Do not allow the nozzle to contact the work.

14. When the cut is complete, release the switch. Shielding gas will continue to flow to cool the torch.

15. Make additional cuts along the remaining lines.

16. Inspect the cut edge of the steel. The edge should be fairly smooth. The cut line should be straight. Work to maintain a constant forward speed while cutting. If dross is present on the metal, adjust your travel speed or current. Make adjustments and make additional cuts. Show your final two cut pieces to your instructor.

## Inspection:

Cuts should be straight with fairly smooth surfaces on the cut edge. There should be no dross on the metal.

Name: _____     Date: _____

Class: _____     Instructor: _____

Lesson Grade: _____     Instructor's Initials: _____

## Assigned Job 23-3
# Plasma Arc Piercing and Cutting Mild Steel

## Objective:

In this job, you will learn to use the plasma arc cutting equipment to pierce and make straight cuts on mild steel. Your instructor will assign you plasma arc cutting equipment to use for this job. Your instructor will also select the type of plasma gas, shielding gas, and electrode you will be using.

**Note**
Do not attempt this job until you have read all safety precautions, satisfactorily completed the *Plasma Arc Cutting (PAC) Safety Test*, and been approved by your instructor.

1. Obtain two pieces of mild steel 1/4″ × 6″ × 6″ (6.4 mm × 150 mm × 150 mm).

2. Mark lines every 3/8″ (9.5 mm). The lines can be drawn with soapstone or chalk or scratched with a sharp object.

3. Answer the following questions about cutting mild steel with the equipment you have been assigned.

   A. What plasma gas will be used? _____

   B. What shielding gas will be used? _____

   C. Refer to Figure 23-7 in the text. What results are expected: excellent, good, or fair? _____

   D. What type of electrode will be used? _____

4. Refer to Figure 23-10 in the text and answer the following questions about cutting 1/4″ (6.4 mm) mild steel. Use the lower current choice for a slower travel speed.

   A. What current is recommended? _____

   B. What travel speed is recommended? _____

5. Inspect your work area and remove any flammable items.

6. Assemble the equipment and install the electrode into the torch.

7. Connect the workpiece lead to the work.

8. Turn on the welding equipment. Open the gas cylinder valve or the valve to the shop manifold system if shielding gas is used. Adjust the regulator to the correct pressure, often about 65 psig. Turn on the water supply if used.

9. Set the correct current on the power supply. If you are not sure of the exact setting, set the current slightly high to ensure a complete cut.

10. Position the torch along the first line drawn, 1/2″ (13 mm) in from the edge. The start of the cut will be a piercing operation. Hold the torch 1/8″–3/16″ (3.0 mm–5.0 mm) from the surface of the work.

11. Hold the torch to create a 60° angle pointing away from you and other people.

12. Press the switch to start the cut. The arc will start and rapidly melt the base metal. Initially, metal will fly upward and away from you. As the pierce forms, rotate the torch to perpendicular and lower it to 1/16″–1/8″ (1.5 mm–3.0 mm) from the surface of the work.

13. Start moving forward. See the recommended travel speed recorded in step 4.

14. Move along the line on the base metal. Do not allow the nozzle to contact the work.

15. When the cut is complete, release the switch. Shielding gas will continue to flow to cool the torch.

16. Make an additional pierce and cut along each of the remaining lines.

17. Inspect the cut edge of the steel. The edge should be fairly smooth. The cut line should be straight. Work to maintain a constant forward speed while cutting. If dross is present on the metal, adjust your travel speed or current. Make adjustments and make additional cuts.

18. When complete, your piece of metal will look like a large comb. There will be many cuts, but all pieces are still attached by a 1/2″ (13 mm) web.

19. Repeat the exercise with your second piece.

20. Show your second piece to your instructor.

## Inspection:

Cuts should be straight with fairly smooth surfaces on the cut edge. The initial pierce may be slightly larger in diameter than the cut line. There should be no dross on the top or bottom surface of the metal.

# Section 6
## Oxyfuel Gas Processes

## Oxyfuel Gas Cutting and Welding Safety Test

### Objective:

You will be able to discuss the safety precautions to be followed when assembling and using oxyfuel gas cutting and welding equipment.

### Instructions:

Always follow safe practices when welding. If you have safety questions or concerns, ask your instructor. This test does not include questions about every safety topic, but is intended to highlight key items. Read all of Chapter 25 and study the Figures 25-1 through 25-25. Also, review the information about oxyfuel gas cutting and welding equipment and protective clothing presented in Chapter 24. Then, answer or complete the following questions.

_____   1. Where should the welder stand when the oxygen or fuel gas cylinder is opened?
      A.  Behind the regulator and gauges.
      B.  In front of the regulators and gauges, to see them better.
      C.  To one side of the regulator and gauges.
      D.  Away from the torch.
      E.  Anywhere near the welding outfit.

2. Where must the torch be at all times while it is lit?

_____

_____   3. The fuse plugs on an acetylene cylinder will normally melt at a temperature of _____ to release the acetylene.
      A.  100°F (38°C)
      B.  212°F (100°C)
      C.  2400°F (1315°C)
      D.  675°F (357°C)
      E.  180°F (82°C)

4. When cylinders are moved or stored, they must have a safety _____ screwed in place.

_____

*For Questions 5–9, the five steps required to shut down an oxyacetylene cutting or welding outfit are listed below, but are not in order. Put them in the correct order by writing the numbers 1–5 in the blanks.*

_____    5.   Completely close the acetylene and oxygen cylinder valves.

_____    6.   Close both regulators by turning the regulator adjusting screws out until they feel loose.

_____    7.   Open both torch valves. This allows all the gases in the system to escape.

_____    8.   Turn off the flame. Close the acetylene torch valve, and then the oxygen torch valve.

_____    9.   After all the pressure gauges read zero, close both torch valves.

*For Questions 10–20, the 11 steps required to properly turn on an oxyacetylene gas cutting or welding outfit are listed below, but are not in the correct order. Put them in the correct order by writing the numbers 1–11 in the blanks.*

_____    10.  Make certain the regulators are closed before opening the cylinder valves. This prevents damage to the regulators and gauges. This is done by turning the regulator adjusting screws counterclockwise on both the oxygen and acetylene regulators. Continue to turn them counterclockwise (outward) until the adjusting screws feel loose in the regulator threads.

_____    11.  Slowly open the acetylene cylinder valve by turning it counterclockwise about 1/4 to 1/2 turn. This provides enough acetylene for most purposes. Use the proper size wrench for the cylinder valve. Leave the wrench in place so that the cylinder can be turned off quickly in an emergency.

_____    12.  Open the acetylene torch valve one-half to one complete turn counterclockwise.

_____    13.  Visually check the torch, valves, hoses, fittings, regulators, gauges, and cylinders for damage.

_____    14.  Close the oxygen torch valve. Check the oxygen regulator for leaks.

_____    15.  Close the acetylene torch valve. Check the acetylene regulator for possible leaks.

_____    16.  Slowly open the oxygen cylinder valve. Remember that pressure in this cylinder can be up to 2200 psig (15.17 MPa). A rapid flow of this high-pressure gas could rupture the regulator diaphragm or gauges. Continue to turn the cylinder valve counterclockwise until it is fully open. The oxygen cylinder valve may leak if it is not fully opened.

_____    17.  Open the oxygen torch valve one-half to one turn counterclockwise.

_____    18.  Turn in (clockwise) the acetylene regulator screw until the working pressure gauge reads the correct working pressure.

_____    19.  Stand to one side of the regulator while you open the cylinder valves. A regulator or gauge that bursts could cause severe injury.

_____    20.  Turn in (clockwise) the oxygen regulator adjusting screw until the correct working pressure is set. Note: When turning on an oxyacetylene cutting outfit, the cutting torch oxygen valve (lever) should be open while the oxygen cutting pressure is set. After setting the working pressure, close the torch valve and release the oxygen cutting lever.

Name _____

_____ 21. When opening a cylinder valve, the regulator adjusting screw should be closed by turning it _____ and the cylinder valve opened _____ to prevent the gauges from rupturing or bursting.
   A. out, slowly
   B. in, slowly
   C. out, quickly
   D. in, quickly

22. *True or False?* Regulator and hose nuts should be tightened with pliers so they are snug and leakproof, but they should not be overtightened.

_____ 23. Acetylene as a pure gas is explosive at pressures above _____ psig (103.4 kPa).

24. If the working pressure gauge shows a drop in pressure after the working pressure is set and the torch valves are closed, something must be leaking. List two things that may be leaking.

_____

_____

_____ 25. What number filter lens should be used in welding goggles for oxyfuel gas welding?
   A. #2–#3.
   B. #4–#6.
   C. #8–#10.
   D. #10–#12.
   E. #12–#14.

_____ 26. The oxygen and fuel gas cylinder openings should be cleaned before the regulators are attached. Which of the following is *not* one of the steps required for this task?
   A. Point the opening away from all workers in the area.
   B. Quickly and briefly open the cylinder valve.
   C. Make sure there are no flames or sparks in the area.
   D. Apply a thin coat of grease to the regulator fittings to prevent thread seizure.

27. Name two items that should never be carried in shirt or trouser pockets while welding.

_____

28. *True or False?* The regulator valves should be opened before the cylinder valves are opened. This prevents a dangerous buildup of pressure in the regulator.

_____ 29. Oxygen and acetylene cylinders must be secured to a wall, column, or portable hand truck. Which of the following should be used?
   A. Wire.
   B. Chain.
   C. Rope.
   D. Any of the above.
   E. None of the above.

30. If the high-pressure gauge loses pressure after the cylinder is turned on, and the regulator adjusting screw is turned all the way out, what may be the cause?

_____

_____

31. *True or False?* After setting the correct oxygen and acetylene working pressures, a leaking regulator valve (needle and seat) is indicated if the working pressure gauge continues to rise.

32. What should be done if a leaking regulator valve is found?

    _____

33. *True or False?* The regulator is opened by turning out the regulator adjusting screw until it feels loose in its threads.

34. Describe the difference between oxygen hoses and fittings and fuel gas hoses and fittings.

    _____

    _____

35. Why are matches never used to light an oxyfuel gas flame? Write your answer in the space provided.

    _____

    _____

    _____

36. *True or False?* The wrench used to open and close the acetylene cylinder valve should be left in place on the valve at all times that the cylinder is in use.

_____    37. About how far should the torch acetylene valve be opened before you light the flame of a welding or cutting torch?
    A.  1/16 turn.
    B.  1/8–1/4 turn.
    C.  1/4–1/2 turn.
    D.  One full turn.
    E.  All the way open.

38. Petroleum-based products should never be used to lubricate the fittings on an oxyacetylene outfit. Name three products that may be petroleum-based.

    _____

    _____

    _____

39. What type of wrench should be used to tighten or loosen the hex nuts on the regulator, torch, and hoses?

    _____

## Lesson 24
# Oxyfuel Gas Cutting and Welding: Equipment and Supplies

### Objectives:

You will be able to describe the gases used in oxyfuel gas cutting and welding, and explain how these gases are stored and distributed. You will be able to explain how regulators, gauges, check valves, flashback arrestors, hoses, and torches operate. You will also know the proper clothing to be worn while welding and be able to select the proper welding goggles.

### Instructions:

Read Chapter 24 and study Figures 24-1 through 24-41. Then, answer or complete the following questions.

_____    1.  The tubing and piping in an acetylene manifold should never be made of ____ because acetylene flowing in pipes made of this material may form explosive compounds.
   A.  steel
   B.  iron
   C.  stainless steel
   D.  copper
   E.  aluminum

2.  When oxygen is stored as a liquid, it is usually kept in a(n) ____ flask.

   _____

3.  Name five fuel gases that have been used successfully in oxyfuel gas welding, soldering, or brazing.

   _____

   _____

_____    4.  A full oxygen cylinder has a pressure of approximately ____.
   A.  212 psig (1500 kPa)
   B.  250 psig (1700 kPa)
   C.  500 psig (3400 kPa)
   D.  2200 psig (15,200 kPa)
   E.  4000 psig (27,600 kPa)

5.  A(n) ____ safely supplies gas from several cylinders.

   _____

6.  A(n) ____ valve keeps oxygen from leaking around the valve stem.

   _____

7. A(n) _____ _____ in the cylinder valve will burst at a predetermined pressure to prevent the oxygen cylinder from exploding in a fire.

_____

8. In case of fire, an acetylene cylinder is equipped with _____ _____, which allow the controlled escape of acetylene gas. This safety feature prevents the cylinder from exploding if overheated.

_____

9. *True or False?* Acetylene becomes very unstable and may burn rapidly or explode at pressures above 15 psig (103 kPa).

10. By filling the cylinders with a porous material and storing the acetylene gas in a liquid called _____, acetylene can be safely stored in cylinders at pressures above the danger point.

_____

11. List five items of clothing that should be worn when welding out of position.

_____

_____  12. If the pressure in a full acetylene cylinder is to be 250 psig (1700 kPa), the highest pressure marked on the cylinder pressure gauge should be at least _____.
   A.  250 psig (1700 kPa)
   B.  375 psig (2600 kPa)
   C.  400 psig (2800 kPa)
   D.  500 psig (3500 kPa)
   E.  750 psig (5200 kPa)

13. Each end of a hose has a fitting. Oxygen fittings have _____-hand threads. Fuel gas fittings have _____-hand threads.

_____

_____  14. To close the regulator and stop the gas flow, turn the regulator adjusting screw in the _____ direction.
   A.  clockwise
   B.  counterclockwise

15. Filter lenses for goggles come in shades from #1 to #14. What shade range of lenses should be worn for oxyfuel gas welding?

_____

_____  16. What type of torch is used with high-pressure acetylene cylinder gases?
   A.  An injector-type torch.
   B.  A positive-pressure torch.

17. When the hole, or orifice, in the end of the torch tip becomes dirty, it may be cleaned using a proper size drill bit or a(n) _____ _____.

18. *True or False?* In a flashback arrestor, the stainless steel filter serves to stop the flame from entering the hose, and the cut-off valve stops the gas flow from the cylinder if a flashback occurs.

_____  19. Why are spark lighters or economizers used instead of matches to light the gases at the torch?
   A.  They are faster.
   B.  They are cleaner.
   C.  They prevent the welder from burning his/her fingers.
   D.  They are cheaper to operate.

20. *True or False?* Pressing the cutting lever on a cutting torch releases a stream of pure acetylene that burns through the preheated metal.

# Lesson 25
# Oxyfuel Gas Cutting and Welding: Equipment Assembly and Adjustment

## Objectives:

You will be able to describe the functions of all parts of the oxyfuel gas cutting and welding outfit. You will also be able to demonstrate how to assemble, turn on, light, and shut down the oxyfuel gas cutting and welding outfit.

## Instructions:

Read Chapter 25 and study Figures 25-1 through 25-25 and then answer or complete the following questions.

1. If the high-pressure gauge continues to drop in pressure after the regulator is turned off, where is the leak?

    _____

2. Name the parts of the oxyfuel gas cutting and welding station shown here.

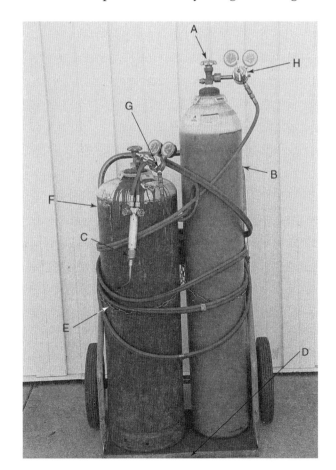

A. _____

B. _____

C. _____

D. _____

E. _____

F. _____

G. _____

H. _____

3. What is the last thing done when you shut down an oxyfuel gas cutting or welding outfit?

_____

_____

_____

_____    4.  A(n) _____ flame has a smooth, bullet-shaped inner cone.
A.  carburizing
B.  neutral
C.  oxidizing
D.  All of the above.

5. Before attaching the regulator, the cylinder _____ must be cleaned to prevent dirt from entering the regulator.

_____

_____    6.  How far should the oxygen cylinder valve be opened for welding and cutting?
A.  1/16–1/8 turn.
B.  1/8–1/4 turn.
C.  1/4–1/2 turn.
D.  1/2–5/8 turn.
E.  All the way.

_____    7.  How far should the acetylene cylinder valve be opened for welding and cutting?
A.  1/16–1/8 turn.
B.  1/8–1/4 turn.
C.  1/4–1/2 turn.
D.  5/8–3/4 turn.
E.  All the way.

8. What type of oxyacetylene flame is shown in the following figure?

_____

9. Describe how hoses and fittings can be tested for leaks.

_____

_____

_____

_____    10.  What type of tool should be used to tighten or loosen the fittings on the regulator, torch, or hoses?
A.  Slip-joint pliers.
B.  Box-end wrench.
C.  Open-end wrench.
D.  Socket wrench.
E.  Pipe wrench.

## Assigned Job 25-1
# Turning On, Lighting, and Shutting Down the Oxyacetylene Outfit

## Objectives:

In this job, you will learn to safely and correctly turn on an oxyacetylene welding outfit equipped with a positive pressure torch. You will also learn how to light and adjust the flame, and how to turn off the outfit.

## Instructions:

Work in a group of three or four students on a cutting or welding outfit assigned by the instructor. The first student to attempt this job will proceed through all procedures (turn on, lighting, adjusting, and shut off), step by step, with group approval before each step is done. When the first student has properly completed all procedures in the job, the next student in the group will repeat the job, once again getting group approval for each step.

Before beginning any step, the student doing the work must tell the group what he or she intends to do. All members of the group must agree that what the student intends to do is correct and is being done in the correct order. The student will then do exactly what is agreed upon. In turn, each student in the group will go through the entire turn on, lighting, adjusting, and shutdown procedures that follow.

> **Note**
> Do not attempt this job until you have read all safety precautions, satisfactorily completed the *Oxyfuel Gas Cutting (OFC) and Welding (OFW) Safety Test*, and been approved by your instructor.

### Turning on the Positive Pressure Oxyacetylene Welding Outfit

1. Examine the welding outfit assigned by the instructor and answer the following questions:
   A. What company manufactured the tip? _____
   B. What is the tip hole (orifice) size? # _____ diameter: _____ in (_____ mm)

2. Determine the correct welding tip orifice size and working pressures for welding 1/8″ (3.2 mm) mild steel. Refer to Figure 24-34 in the text.
   A. Oxygen pressure: _____ psig
   B. Acetylene pressure: _____ psig
   C. Orifice size: _____

3. Determine the next step required to properly turn on the cutting or welding outfit. Inform the group of what you think the next step should be. If *all* members of the group agree, perform the step exactly as agreed upon.

4. Repeat step 3 until the cutting or welding outfit has been properly turned on.

5. Have the instructor approve the work.

*Instructor's initials:* _____

## Lighting and Adjusting the Oxyacetylene Flame

1. Before beginning to light the torch and adjust the flame, describe to the group what you intend to do as the first step. If *all* members of the group agree, perform the task exactly as agreed upon.

2. Determine the next step required to light or adjust the torch. Describe to the group what you think that step should be. If *all* members in the group agree, perform the task exactly as agreed upon.

3. Repeat step 2 until the torch is lit and adjusted to a neutral flame.

4. Have the instructor approve the work.

*Instructor's initials:* _____

## Shutting Down the Oxyacetylene Cutting and Welding Outfit

1. Before beginning to shut down the cutting or welding outfit, describe to the group what you intend to do as the first step. If *all* members of the group agree, perform the task exactly as agreed upon.

2. Determine the next step required to properly shut down the cutting or welding outfit. Describe to the group what you think that step should be. If *all* members of the group agree, perform the step exactly as agreed upon.

3. Repeat step 2 until the outfit has been properly shut down.

4. Have the instructor approve the work.

> **Note**
> Be careful not to trap pressure in the high-pressure side of the regulator and in the high-pressure gauge, which would indicate an improper shutdown procedure.

*Instructor's initials:* _____

## Lesson 26
# Oxyfuel Gas Cutting

## Objectives:

You will be able to demonstrate how to assemble, adjust, and use the equipment required for oxyfuel gas cutting. You will gain skill in cutting metal with the oxyfuel gas cutting process.

## Instructions:

Read Chapter 26 and study Figures 26-1 through 26-15. Also, review the procedures in Chapter 25 for turning on, lighting, adjusting, and shutting down a cutting torch. Then, answer or complete the following questions.

1. *True or False?* When a cutting torch attachment is being used, the oxygen torch valve on the cutting attachment must be opened one full turn.

2. A piece of _____ _____ clamped to the work may be used to achieve a straighter kerf line when cutting manually.

3. What type of slag is unacceptable when cutting metal with the oxyfuel gas process?

   _____

4. What are the suggested oxygen and acetylene gas pressures for cutting 1/2″ (12.7 mm) mild steel when using a positive pressure torch?
   Oxygen pressure: _____ psig
   Acetylene pressure: _____ psig

_____ 5. What causes a bell-mouthed kerf?
   A. Too much acetylene pressure.
   B. Too much oxygen pressure.
   C. Moving too fast.
   D. Moving too slowly.
   E. None of the above.

6. *True or False?* Cutting tip sizes are identified by numbers ranging from 0–10.

7. Name five fuel gases that have been used with oxygen for oxyfuel cutting.

   _____

   _____

   _____

8. A motor-driven carriage keeps _____ and _____ consistent because it moves along a track.

_____     9. Which of the following does not affect the quality of the cut (kerf)?
           A. The cutting tip size.
           B. The torch cutting speed.
           C. A steady torch movement.
           D. The oxygen pressure.
           E. The size of the cutting torch.

10. *True or False?* An electric motor-driven carriage is adjustable for flame height, cutting speed, and torch angle.

11. If a positive pressure cutting torch is equipped with a cutting torch attachment, what additional step must be taken when shutting it down?

_____

12. What is the cutting speed when cutting 2″ (50 mm) thick steel?

_____

_____     13. What can be done if the available cutting tips are too large for the metal thickness being cut?
           A. Decrease the fuel gas pressure.
           B. Increase the oxygen pressure.
           C. Increase the fuel gas pressure.
           D. Decrease the cutting speed.
           E. Decrease the angle between the base metal and the tip.

_____     14. What is the ignition temperature of mild steel?
           A. 2470°F (1343°C).
           B. 816°F (436°C).
           C. 1500°F (816°C).
           D. 2470°C (4478°F).
           E. 1500°C (2732°F).

15. When cutting thick steel, a higher volume of _____ is needed. A high-pressure, high-volume _____ may be required.

16. *True or False?* A minimum of two preheating orifices should line up with the cutting line.

17. Identify the holes (orifices) in the cutting tip shown:

A. _____

B. _____

Name _____

18. Which type cutting machine is used to cut and bevel pipe?

_____

_____

_____

_____ 19. Which of the two cutting tips in the drawing is aligned properly with the cutting line, A or B?

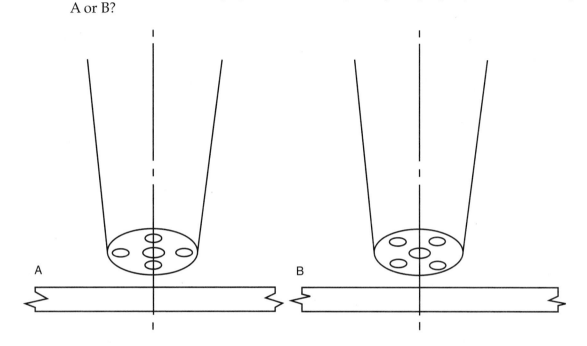

20. How far is the flame from the base metal when cutting using a motor-driven carriage-mounted cutting torch?

_____

## Assigned Job 26-1
# Manually Cutting 1/4"–5/8" (6.4 mm–15.9 mm) Mild Steel

## Objectives:

In this job, you will learn to assemble the cutting outfit and set the correct gas pressures. You will also learn to perform acceptable cuts on 1/4"–5/8" (6.4 mm–15.9 mm) mild steel, using a positive pressure cutting torch or a cutting attachment on a positive pressure welding torch.

> **Note**
> Do not attempt this job until you have read all safety precautions, satisfactorily completed the *Oxyfuel Gas Cutting (OFC) and Welding (OFW) Safety Test*, and been approved by your instructor.

1. Obtain at least four pieces of mild steel that are 1/4"–5/8" (6.4 mm–15.9 mm) thick, at least 6" (150 mm) long, and at least 6" (150 mm) wide. Scrap steel or weld samples may be used.

2. Fill in the following information for the metal thickness being used. Refer to Figure 26-1 in the text.
   A. Recommended tip number: _____
   B. Recommended oxygen pressure: _____ psig
   C. Recommended acetylene pressure: _____ psig
   D. Recommended cutting speed: _____ in/min
   E. Recommended tip angle from vertical: _____

3. Attach a positive pressure cutting torch to the oxygen and acetylene hoses.

4. Using a proper size open-end wrench, check all connections to ensure that they are tight. Do not overtighten brass fittings.

5. Draw lines lengthwise on the metal to be cut. These lines should be 1 1/2" (40 mm) from the edge and 1 1/2" (40 mm) apart. Use a soapstone or chalk to draw the lines.

> **Caution**
> Remove any flammable items from the cutting area before lighting the torch and starting to cut. Hot metal will fly a long way as cutting takes place.

6. Turn on the oxyacetylene cutting outfit. Set the correct pressures for both oxygen and acetylene.

7. Light the preheating flames and adjust to a neutral flame. Turn on the cutting oxygen and check that the preheating flames remain neutral. If they do not, adjust the oxygen torch valve until they remain neutral when the cutting oxygen is turned on.

8. Three cuts are to be made. The cut pieces will be at least 1 1/2″ (40mm) wide and at least 6″ (150 mm) long. These cuts must be made freehand (without the use of an angle iron or other guide).

9. Begin each cut by holding the torch tip at the edge of the metal. Hold the preheating flames about 1/16″–1/8″ (1.5 mm–3.0 mm) from the metal surface and tilt the tip away from vertical to the correct angle. Hold the tip at the edge until the metal turns bright red. The metal is now ready to cut. Press the cutting oxygen lever, and move the tip along the cut line, at the recommended speed, to the end of the cut.

10. Using an angle iron as a guide to make the cut straighter, make three more cuts on another piece of 6″ × 6″ (150 mm × 150 mm) mild steel using the same dimensions as in steps 5 and 8.

11. Shut down the cutting outfit and remove the cutting torch. Connect a cutting attachment to the outfit. Use the same size tip as in step 2. Adjust the preheating flames and repeat steps 7 through 10.

## Inspection:

Each piece must have edges that are square to the surface. The cut line should be relatively straight. There should be no hard slag on the underside of the metal.

Name: _____   Date: _____

Class: _____   Instructor: _____

Lesson Grade: _____   Instructor's Initials: _____

# Assigned Job 26-2
# Manually Cutting Mild Steel Less than 1/8" (3.2 mm) Thick

## Objective:

In this job, you will learn to make acceptable cuts on mild steel that is less than 1/8" (3.2 mm) thick.

> **Note**
>
> Do not attempt this job until you have read all safety precautions, satisfactorily completed the *Oxyfuel Gas Cutting (OFC) and Welding (OFW) Safety Test*, and been approved by your instructor.

1. Obtain a piece of mild steel that measures 1/16" × 6" × 6" (1.6 mm × 150 mm × 150 mm).

2. Draw lines on the surface of the metal 1 1/2" (40 mm) apart.

3. Examine Figure 26-1 in the text. This figure does not list a cutting tip for use on metal less than 1/8" (3.2 mm) thick.

4. Fill in the following information for the 1/8" thickness listed in Figure 26-1:
   A. Recommended tip number: _____
   B. Recommended oxygen pressure: _____ psig
   C. Recommended acetylene pressure: _____ psig
   D. Recommended cutting speed: _____ in/min
   E. Recommended tip angle (from horizontal) for thin metal: _____

5. Start the cutting outfit, set the correct pressures, as listed in step 4, and adjust the preheat flames to a neutral setting. Recheck the flame adjustment after the cutting oxygen valve is opened.

6. Set the base metal to be cut on the cutting table.

7. Hold the preheat flames at the edge of the base metal. Hold the preheat flames about 1/16"–1/8" (1.5 mm–3.0 mm) above the metal surface. The cutting tip must be held at the angle recorded in step 4. By holding the tip at this low angle, the thickness of the metal being cut is increased. This makes it easier for this larger tip to cut the thin metal.

8. Make at least three practice cuts. Each piece of metal that is cut should be 1 1/2" (40 mm) wide.

## Inspection:

The edges of the pieces must be square to the surface. The cutting lines should be relatively straight; there must not be any hard slag on the underside of the pieces.

## Assigned Job 26-3
# Cutting Mild Steel Using a Motorized Carriage and Track

## Objective:

In this job, you will learn to assemble, adjust, and operate a motorized carriage and track to make acceptable cuts on mild steel.

> **Note**
> Do not attempt this job until you have read all safety precautions, satisfactorily completed the *Oxyfuel Gas Cutting (OFC) and Welding (OFW) Safety Test*, and been approved by your instructor.

1. Name the parts of the motorized carriage and torch shown here:

A. _____     E. _____

B. _____     F. _____

C. _____     G. _____

D. _____

2. Obtain a piece of mild steel with a thickness of at least 1/4″ (6.4 mm), a width of at least 6″ (150 mm), and at least 12″ (305 mm) in length.

3. Answer the following questions for the thickness of metal that you will cut. Refer to Figure 26-1 in the text.
   A. Recommended preheat orifice size: _____
   B. Recommended oxygen pressure: _____ psig
   C. Recommended acetylene pressure: _____ psig
   D. Recommended cutting speed: _____ in/min

4. Set up the track on a cutting table so that cuts can be made across the width of the metal.

5. Place the motorized carriage and torch on the track. Connect the oxygen and acetylene hoses to the cutting torch. Check all connections to ensure that they are tight.

6. Open the oxygen and acetylene cylinders and set the correct pressures on the regulators.

7. Move the carriage and torch away from the metal to be cut and light the preheating flames. Adjust the flames to neutral in the same way you would for a positive pressure cutting torch.

8. Adjust the torch angle, height, and cutting speed. Move the carriage and torch to the edge of the metal to be cut.

9. Heat the edge of the metal to a bright red, turn on the cutting oxygen, and engage the carriage clutch for forward motion. At the end of the cut, disengage the carriage clutch so the carriage stops.

10. Make a practice cut. If the cut is not acceptable, adjust the oxygen pressure, flame height, torch angle, or cutting speed to obtain an acceptable cut.

11. Make at least three practice cuts. Each piece of metal that is cut should be 1 1/2″ (40 mm) wide.

## Inspection:

Each piece must have square edges and a straight, clean kerf. There should be no hard slag on the underside of the metal.

## Lesson 27
# Oxyfuel Gas Welding: Flat Welding Position

## Objectives:

You will be able to demonstrate how to oxyfuel gas weld an outside corner joint, lap joint, butt joint, and T-joint in the flat welding position.

## Instructions:

Read Chapter 27 and study Figures 27-1 through 27-19. Also, review Chapter 7 for information about welding joints and welding positions. Then, answer or complete the following questions.

_____    1.  What conditions cause a welding rod to "freeze" in the weld pool?
  A.  Welding rod too small.
  B.  Welding tip too small or welding rod too large in diameter.
  C.  Welding rod too small and the welding tip too large.
  D.  Welding tip too large.
  E.  None of the above.

2.  Generally, what size leg should a fillet weld have for the thicknesses of mild steel listed?
  A. 1/8" (3.2 mm): _____
  B. 1/4" (6.4 mm): _____
  C. 1/2" (12.7 mm): _____

3.  Two angles are used to describe the position of a welding torch as a weld is created. These angles are the ____ angle and the ____ angle.

  _____

4.  What causes the appearance of a "keyhole" at the root of a bevel or V-groove butt weld?

  _____

  _____

  _____

5.  When making a fillet weld, what does the welder look for to tell him or her that the surface has melted and the welding rod can be dipped into the weld pool?

  _____

_____     6.   The torch flame is pointed more toward the _____ piece when welding a lap joint.
           A.   surface
           B.   edge
           C.   vertical
           D.   horizontal

   7.  What melts the tip of the welding rod during oxyfuel gas welding?

   _____

   8.  Write the technical definition of work angle.

   _____

   _____

   9.  Write the technical definition of travel angle.

   _____

   _____

  10.  When welding an inside corner or T-joint, what is the recommended work angle? _____

_____    11.   On practice pieces, tack welds are normally placed _____ apart.
           A.   2″
           B.   3″
           C.   4″
           D.   5″
           E.   6″–9″

_____    12.   At what angle to the base metal is the welding rod held when welding a butt, lap, or
                 inside corner joint in the flat welding position?
           A.   5°–10°.
           B.   15°–45°.
           C.   30°–60°.
           D.   60°–90°.
           E.   90°.

_____    13.   At what travel angle is the torch held when laying a weld bead or butt welding in the
                 flat welding position?
           A.   0°
           B.   10°–15°.
           C.   15°–20°.
           D.   35°–45°.
           E.   45°–60°.

_____    14.   Which of the following conditions would not cause a weld pool to create a hole in the
                 base metal?
           A.   The weld pool gets too wide.
           B.   The weld pool gets too deep.
           C.   The torch tip is too large.
           D.   The forward travel speed is too fast.

Name _____

_____ 15. Which method of welding is being done in the following illustration?
A.  Forehand.
B.  Backhand.

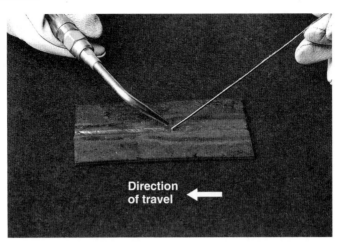

16. *True or False?* A right-handed welder holds the torch in the right hand and the welding rod in the left hand.

17. The two ways of holding a welding torch are to hold it like a(n) _____ or like a(n) _____.

18. Name the weld defects shown in the following drawing:

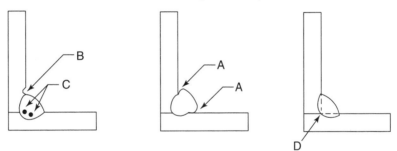

A. _____          C. _____

B. _____          D. _____

19. A work angle of 0° can also be described as an angle of _____ from the base metal.

20. Explain why the metal in a butt joint is tack welded before it is welded.

_____

_____

Name: _____  Date: _____

Class: _____  Instructor: _____

Lesson Grade: _____  Instructor's Initials: _____

# Creating a Continuous Weld Pool on Mild Steel

## Objective:

In this job, you will use oxyacetylene welding to create a continuous weld pool on mild steel. You will learn to control the molten metal and penetration without melting through the other side.

> **Note**
> Do not attempt this job until you have read all safety precautions, satisfactorily completed the *Oxyfuel Gas Cutting (OFC) and Welding (OFW) Safety Test*, and been approved by your instructor.

1. Obtain one piece of 1/16″ (1.6 mm) or thinner mild steel measuring 4″ × 6″ (100 mm × 150 mm).

2. Draw four lines 6″ (150 mm) long and 1″ (25 mm) apart on the metal surface, using a soapstone or chalk. The first line should be 1/2″ (13 mm) from the edge. See the following drawing.

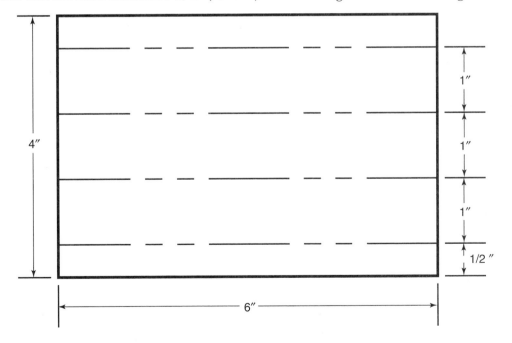

3. Fill in the following information for your torch, tip, and base metal thickness. Refer to Figure 24-34 in the text.
   A. The manufacturer of your positive pressure torch: _____
   B. Drill size of tip: #_____
   C. Manufacturer's number size of tip: #_____
   D. Oxygen working pressure: _____ psig
   E. Acetylene working pressure: _____ psig

4. Place the metal across two firebricks on your welding table, so that the outer edges are supported by the bricks.

5. Turn on the outfit, set the working pressures, light the flame, and adjust to a neutral flame.

6. This weld pool exercise should be completed with the forehand welding method. If you are right-handed, start at the right side of the metal and proceed to your left. If you are left-handed, start at the left side of the metal and proceed to your right.

7. Point the flame in the direction of travel. The tip of the flame must be kept about 1/16"–1/8" (1.5 mm–3.0 mm) above the metal surface. This produces the maximum heat input from the torch. The distance can be varied up and down to control the heat input to the metal. The torch tip should be held at a 0° work angle and a 35°–45° travel angle. See Figure 27-3 in the text. The travel angle may increase or decrease to control the heat input.

8. Create a round weld pool on the metal and watch it carefully. When it sags downward, move ahead slightly, but note the width of the pool at the time it sagged. This is the width you should maintain as you move the weld pool ahead. By maintaining the proper weld pool width, you can obtain a uniform amount of penetration on the underside of the base metal.

9. Carry one weld pool along the entire length of the first line. Use pliers to pick up the hot metal and inspect the bottom side for a good, even penetration.

10. Practice carrying a continuous weld pool along the other three lines on the plate. Keep your weld pool lines straight across the length of the metal.

11. Inspect the plate after each weld pool is advanced across it. If necessary, correct the tip selection, travel speed, or weld pool width so that the next weld bead is better than the last.

12. If necessary, try again with another piece of base metal until four good-quality weld beads are produced on one piece.

## Assigned Job 27-2
# Welding without a Welding Rod

## Objective:

In this job, you will learn to make a weld on a square-groove outside corner butt joint without a welding rod.

> **Note**
>
> Do not attempt this job until you have read all safety precautions, satisfactorily completed the *Oxyfuel Gas Cutting (OFC) and Welding (OFW) Safety Test*, and been approved by your instructor.

1.  Obtain eight pieces of 1/16″ (1.6 mm) low-carbon steel, each measuring 1 1/2″ × 6″ (40 mm × 150 mm).

2.  Fill in the following information for your torch, tip, and base metal thickness. Refer to Figure 24-34 in the text.
    A. Drill size of tip: #_____
    B. Tip number size: #_____
    C. Oxygen working pressure: _____ psig
    D. Acetylene working pressure: _____ psig

3.  Arrange the pieces as shown in the following illustration. The vertical piece should overlap the horizontal piece by the thickness of the base metal. The weld axis should run from right to left. The pieces may be supported by a firebrick.

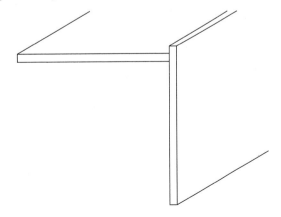

4.  Turn on the welding outfit, set the working pressures, light the torch, and adjust the flame to neutral.

5. Tack weld the metal in three places (beginning, middle, and end). This is done by heating the vertical overlapped edge until it melts into the horizontal piece.

6. Start the weld at the end of the weldment closest to your welding hand and make a forehand weld. Hold the torch at a 0° work angle and a 35°–45° travel angle. Point the torch flame at the vertical edge and melt it down into the horizontal piece. The weld bead should be about three times as wide as the metal is thick. A crescent or zigzag motion may be used to move the molten metal from the vertical piece into the horizontal weld pool. Complete fusion must occur. A small amount of penetration should be seen on the inside corner of the joint.

7. Use four pieces of metal to make two practice welds like the one described in the previous step.

8. For a grade, use the last four pieces of metal and make two additional welds like the one described in step 5.

## Inspection:

Each weld bead should be slightly convex and have an even width and evenly spaced ripples. A small amount of penetration should show on the inside corner of the joint.

## Assigned Job 27-3
# Making Stringer and Weave Beads

## Objectives:

In this job, you will learn to make stringer beads and weave beads using filler metal.

---

**Note**

Do not attempt this job until you have read all safety precautions, satisfactorily completed the *Oxyfuel Gas Cutting (OFC) and Welding (OFW) Safety Test*, and been approved by your instructor.

---

1. Obtain two pieces of 1/8″ (3.2 mm) low-carbon steel, each measuring 4″ × 6″ (100 mm × 150 mm). Also obtain a mild steel filler rod of the correct size.

2. Draw four straight lines on the metal surface with soapstone or chalk. The lines should be 1″ (25 mm) apart and 1/2″ (13 mm) from the edge. See the following drawing.

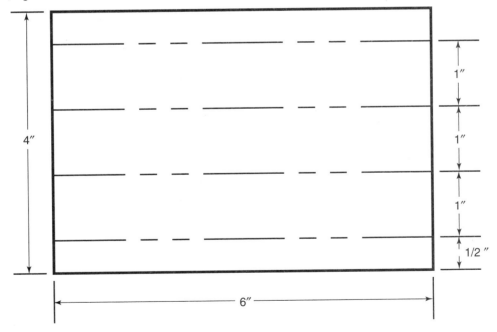

3. Place the metal on a firebrick on your workbench.

4. Fill in the following information for your torch, tip, and base metal thickness. Refer to Figure 24-34 in the text.
   A. Correct drill size for tip: _____
   B. Tip number size: #_____
   C. Oxygen working pressure: _____ psig
   D. Acetylene working pressure: _____ psig
   E. Suggested welding rod diameter: _____
   F. Suggested work angle: _____, which is an angle of _____ off the surface.
   G. Suggested travel angle: _____, which is an angle of _____ off the surface.
   H. Suggested welding rod angle from the base metal: _____
   I. Distance of flame tip from the weld pool surface: _____

5. Turn on your welding station, set the working pressures, and then light the torch and set the flame to neutral.

6. These weld beads will be made using forehand welding. Begin by creating a weld pool that is about two times as wide as the metal is thick. Hold the torch at a 0° work angle and a 35°–45° travel angle. Hold the welding rod at a 15°–45° angle from the base metal. Dip the welding rod into the leading edge of the pool. Withdraw the welding rod and move the flame and weld pool forward slightly. Dip the rod again. Dip the rod as often as needed to create a slightly convex weld bead. The weld bead should be a stringer bead. Do not move the flame from side to side. Continue this process until the weld bead is completed along the 6″ (150 mm) length.

7. Make a number of practice stringer beads.

8. Next, make several weave beads. As the weld moves along the weld axis, move the flame across the axis about 1/4″ (6.0 mm) on each side. The completed weave bead should be about 1/2″ (13 mm) wide.

9. After practicing, make two stringer beads and two weave beads on the second piece of metal. This second piece will be graded by your instructor.

## Inspection:

All welds should have a uniform width, have a slightly convex weld bead with uniform ripples, and be straight along the weld axis. No penetration is required in this exercise.

## Assigned Job 27-4
# Welding a Lap Joint in the Flat Welding Position

## Objective:
In this job, you will learn to make a fillet weld on a lap joint in the flat welding position.

> **Note**
> Do not attempt this job until you have read all safety precautions, satisfactorily completed the *Oxyfuel Gas Cutting (OFC) and Welding (OFW) Safety Test*, and been approved by your instructor.

1. Obtain four pieces of 1/16″ (1.6 mm) low-carbon steel, each measuring 1 1/2″ × 6″ (40 mm × 150 mm), and four pieces of 1/8″ (3.2 mm) low-carbon steel, each measuring 1 1/2″ × 6″ (40 mm × 150 mm). Also obtain two welding rods of the correct size for each metal thickness.

2. Fill in the following information for your torch, tip, and metal thickness:

|  | 1/16″ | 1/8″ |
|---|---|---|
| A  Correct tip drill size: | ____ | ____ |
| B.  Tip number size: | ____ | ____ |
| C.  Oxygen working pressure: | ____ psig | ____ psig |
| D.  Acetylene working pressure: | ____ psig | ____ psig |
| E.  Welding rod diameter: | ____ | ____ |
| F.  Fillet weld leg size: | ____ | ____ |

3. What are the suggested welding angles and distances for making lap joints?
   A. Suggested work angle: ____ (which is an angle of ____ off the surface).
   B. Suggested travel angle: ____ (which is an angle of ____ off the surface).
   C. Where should the flame be directed?

   _____

   D. How far from the weld pool surface should the flame be? _____
   E. What is the correct angle for the welding rod from the base metal? _____

4. Turn on the welding outfit and set the working pressures. Light the torch and adjust the flame to neutral.

5. Place two pieces of 1/16″ (1.6 mm) metal on a firebrick and position the two pieces on top of each other so that they overlap about 3/4″ (19 mm). Tack weld each side of the lap joint three times.

6. Support the tacked weldment with a second firebrick or metal block so that the weld face and weld axis are both horizontal.

7. Make a practice fillet weld on both sides of the 1/16″ (1.6 mm) metal, as shown by the AWS welding symbols on the following drawing. Turn the weldment as needed to make both welds in the flat welding position.

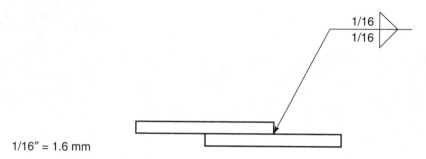

1/16″ = 1.6 mm

8. Tack weld two pieces of 1/8″ (3.2 mm) metal and make practice fillet welds on both sides, as shown in the following drawing. Turn the weldment as needed to make both welds in the flat welding position.

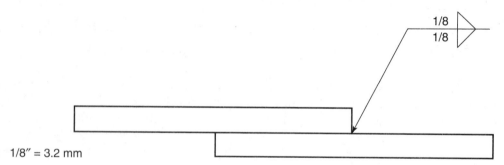

1/8″ = 3.2 mm

9. Repeat steps 5–8 with the remaining four pieces of metal, this time for a grade.

## Inspection:

Each fillet weld should be straight and even in width. The weld bead should be slightly convex and have evenly spaced ripples. The weld bead should not be undercut or overlapped.

Name: _____    Date: _____

Class: _____    Instructor: _____

Lesson Grade: _____    Instructor's Initials: _____

## Assigned Job 27-5
# Welding a T-Joint in the Flat Welding Position

## Objective:

In this job, you will learn to make a fillet weld on a T-joint in the flat welding position.

> **Note**
>
> Do not attempt this job until you have read all safety precautions, satisfactorily completed the *Oxyfuel Gas Cutting (OFC) and Welding (OFW) Safety Test*, and been approved by your instructor.

1. Obtain four pieces of 1/16″ (1.6 mm) low-carbon steel, each measuring 1 1/2″ × 6″ (40 mm × 150 mm), and two pieces of 1/16″ (1.6 mm) low-carbon steel, each measuring 3″ × 6″ (75 mm × 150 mm). Also obtain two welding rods of the correct size.

2. Fill in the following information for your torch, tip, and metal thickness:
   A. Suggested tip drill size: _____
   B. Tip number size: _____
   C. Oxygen working pressure: _____ psig
   D. Acetylene working pressure: _____ psig
   E. Welding rod diameter: _____
   F. Fillet leg size: _____

3. What are the suggested welding angles and distances for T-joints?
   A. Suggested work angle: _____ (which is an angle of _____ off the surface).
   B. Suggested travel angle: _____ (which is an angle of _____ off the surface).
   C. How far from the base metal should the flame be held? _____
   D. At what angle is the welding rod held to the base metal? _____

4. Turn on the welding outfit, set the working pressures, light the flame, and adjust it to neutral.

5. As a practice piece, assemble the base metal as shown in the following drawing. Tack weld each joint in three places on each side. Turn the weldment as needed to make all welds in the flat welding position.

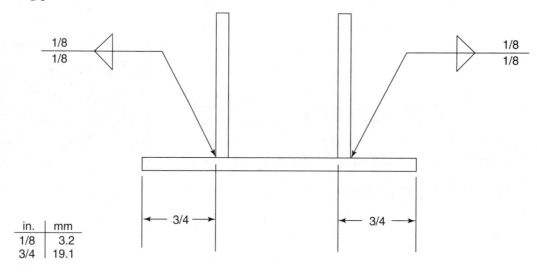

| in. | mm |
|-----|------|
| 1/8 | 3.2 |
| 3/4 | 19.1 |

6. Complete the welds as shown in the drawing. In making the fillet welds on the T-joints in between the two vertical pieces, the torch angles and the method of holding the torch may need to be adjusted. Many welded joints present a challenge to be accomplished.

7. Repeat steps 5 and 6, using the remaining pieces of metal. Present the completed weldment to your instructor for a grade.

## Inspection:

Each fillet weld should be inspected before welding the next joint. Each weld bead should be even in width, slightly convex, straight, and have evenly spaced ripples. The weld beads should not be undercut or overlapped.

## Assigned Job 27-6
# Welding a Square-Groove Butt Joint in the Flat Welding Position

### Objective:

In this job, you will learn to weld a square-groove butt joint in the flat welding position. You will use a stringer bead to make this weld.

> **Note**
> Do not attempt this job until you have read all safety precautions, satisfactorily completed the *Oxyfuel Gas Cutting (OFC) and Welding (OFW) Safety Test*, and been approved by your instructor.

1. You will need six pieces of 1/16″ (1.6 mm) mild steel, each measuring 1 1/2″ × 6″ (40 mm × 150 mm) and two welding rods of the correct diameter.

2. Fill in the following information for your torch, tip, and base metal thickness:
   A. Correct tip drill size: _____
   B. Tip number size: _____
   C. Oxygen working pressure: _____ psig
   D. Acetylene working pressure: _____ psig
   E. Welding rod diameter: _____

3. What are the suggested welding angles and distances for this job?
   A. What is the suggested work angle? _____ (which is _____ off the surface).
   B. What is the suggested travel angle? _____ (which is _____ off the surface).
   C. How far from the base metal is the torch flame held? _____
   D. At what angle to the weld axis is the welding rod held? _____

4. Turn on the welding outfit, set the working pressures, light the torch, and adjust the flame to neutral.

5.  Place the two pieces of metal on a firebrick. To maintain the correct spacing, bend a 1/16″ diameter filler rod into a U shape and place it between the pieces, as shown in the following illustration. This ensures the proper spacing of about 1/16″ (1.6 mm) along the entire length of the joint. Tack weld the pieces in three places to form the joint shown in the following drawing. This weldment will be used for practice. Remove the U-shaped spacer after the tack welding is completed.

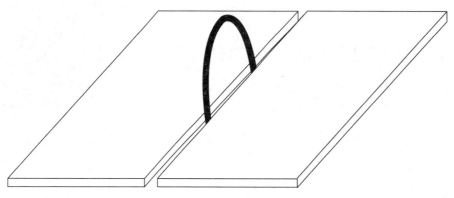

**Note**
A space of about 1/16″ (1.6 mm) should exist between the pieces after the tack welding is completed.

6.  Complete the butt weld as shown by the AWS welding symbol in the drawing.

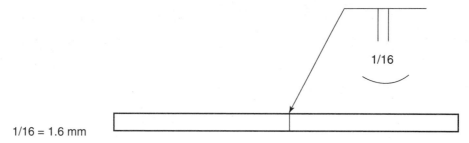

1/16 = 1.6 mm

7.  Repeat steps 5 and 6 with the other four pieces of metal. These last two weldments will be used for a grade. All welds are to be made in the flat welding position.

## Inspection:

Each butt weld should have a small and uniform amount of penetration showing on the side opposite the weld bead face. Each weld bead is to be straight and slightly convex, have an even width and evenly spaced ripples, and have no visible inclusions.

## Lesson 28
# Oxyfuel Gas Welding:
# Horizontal, Vertical, and Overhead Welding Positions

## Objectives:

You will be able to demonstrate how to use oxyfuel gas welding to weld lap, outside corner, T-, and butt joints in the horizontal, vertical, and overhead welding positions.

## Instructions:

Read Chapter 28 and study Figures 28-1 through 28-26. Then, answer or complete the following questions.

1. *True or False?* The same travel angles and work angles are used to weld a butt joint in any position.

2. For an overhead fillet weld on a T-joint, what are the correct angles for A, B, and C in the following drawing?

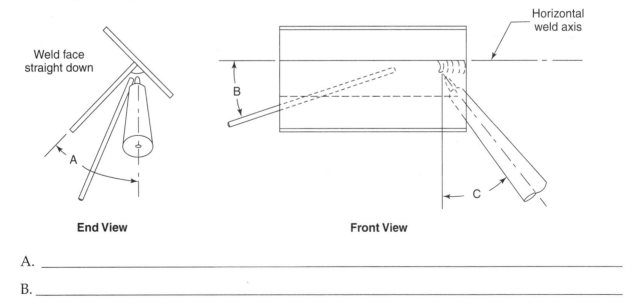

**End View**          **Front View**

A. _____

B. _____

C. _____

3. When making a horizontal butt weld, the torch is held at a(n) _____ work angle and a(n) _____ travel angle to keep the metal in the weld pool and the weld bead from sagging downward.

4. *True or False?* Welds made out of position by a qualified welder will be as strong and look as good as welds made in the flat welding position.

5. In the following drawing of an outside corner joint being welded in the vertical welding position, what are the correct angles for A and B?

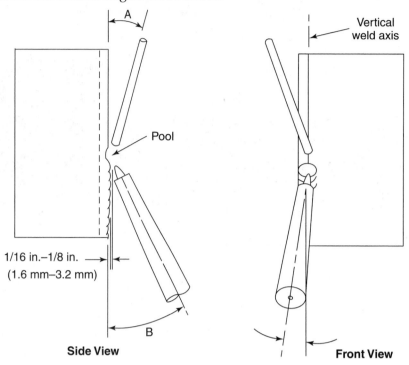

**Side View**                    **Front View**

A. _____

B. _____

6. When making a fillet weld on a lap joint, what should be the shape of the weld pool?

_____ 7. The weld axis must be _____ to be considered a horizontal weld.
   A. within 5° of horizontal
   B. within 10° of horizontal
   C. within 15° of horizontal
   D. within 20° of horizontal
   E. within 30° of horizontal

8. When you are practicing the horizontal fillet weld on a lap joint, the weld face should be at an angle of _____ or _____.

_____

_____ 9. When the face of the fillet weld on a lap, inside corner, or T-joint is between 90°–140°, the weld is being made in which welding position?
   A. Flat.
   B. Horizontal.
   C. Vertical.
   D. Overhead.
   E. The welding position is completely independent of the position of the weld face.

_____ 10. Welds made in the overhead position are usually made with a _____.
   A. forehand motion
   B. backhand motion

Name: _____     Date: _____

Class: _____     Instructor: _____

Lesson Grade: _____     Instructor's Initials: _____

## Assigned Job 28-1
# Welding a Lap Joint in the Horizontal Welding Position

## Objective:

In this job, you will learn to make a fillet weld on a lap joint in the horizontal welding position.

**Note**

Do not attempt this job until you have read all safety precautions, satisfactorily completed the *Oxyfuel Gas Cutting (OFC) and Welding (OFW) Safety Test*, and been approved by your instructor.

1. Obtain five pieces of low-carbon steel that measure 1/16″ × 1 1/2″ × 6″ (1.6 mm × 40 mm × 150 mm) and five pieces that measure 1/8″ × 1 1/2″ × 6″ (3.2 mm × 40 mm × 150 mm). Also obtain two welding rods of the correct size for each metal thickness.

2. Fill in the following information for your torch, tip, and metal thickness. Refer to Figure 24-34 in the text.

|  | 1/16″ | 1/8″ |
|---|---|---|
| A. Correct tip drill size: | _____ | _____ |
| B. Tip number size: | _____ | _____ |
| C. Oxygen working pressure: | _____ psig | _____ psig |
| D. Acetylene working pressure: | _____ psig | _____ psig |
| E. Welding rod diameter: | _____ | _____ |
| F. Fillet weld leg size (see drawings for steps 10 and 11): | _____ | _____ |

3. What are the suggested welding angles and distances for this job?
   A. Suggested work angle: _____ (which is an angle of _____ off the surface).
   B. Suggested travel angle: _____ (which is an angle of _____ off the surface).
   C. How far from the base metal is the flame held? _____
   D. At what angle is the welding rod held to the base metal? _____

4. Turn on the welding outfit, set the working pressures, and adjust the flame to neutral.

5. Place two pieces of 1/16″ (1.6 mm) metal on a firebrick. The metal should be overlapped 3/4″ (19 mm). Tack weld each side of the lap joint in three places. This may be done in the flat welding position.

6. Repeat step 5, using two pieces of 1/8″ (3.2 mm) metal.

7. Place the 1/16″ metal weldment into a weld positioner. Weld a 1/16″ × 1/8″ (1.6 mm × 3.2 mm) fillet weld on both sides of the weldment. These welds must be made in the horizontal welding position.

8. Repeat step 7, using the 1/8″ metal weldment. The fillet welds on this weldment must have 1/8″ (3 mm) legs.

9. When making these welds, the weld face must be 135° or 225° from the straight-down position. The weld axis should be within 15° of horizontal.

10. Tack weld the remaining three pieces of 1/16″ (1.6 mm) metal to form the weldment shown. Tack welding may be done in the flat welding position.

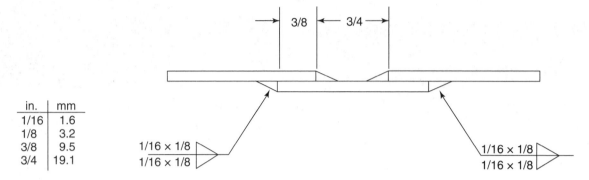

| in. | mm |
|-----|------|
| 1/16 | 1.6 |
| 1/8 | 3.2 |
| 3/8 | 9.5 |
| 3/4 | 19.1 |

11. Tack weld the 1/8″ (3.2 mm) metal as shown in the following drawing.

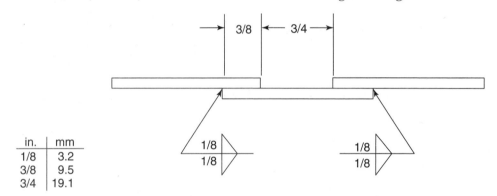

| in. | mm |
|-----|------|
| 1/8 | 3.2 |
| 3/8 | 9.5 |
| 3/4 | 19.1 |

12. Refer to the drawing in step 10 and make the fillet welds on the 1/16″ (1.6 mm) metal as directed by the AWS welding symbols. Turn the weldment as needed, so all welds can be made in the horizontal welding position. You will be graded on these welds.

13. Refer to the drawing in step 11 and make the fillet welds on the 1/8″ (3.2 mm) metal as directed by the AWS welding symbols. Turn the weldment as needed, so all welds can be made in the horizontal welding position. You will also be graded on these welds.

## Inspection:

Each fillet weld should be straight and slightly convex, with an even width and evenly spaced ripples in the weld bead. There should be no undercut or overlap present.

Name: _____  Date: _____

Class: _____  Instructor: _____

Lesson Grade: _____  Instructor's Initials: _____

# Welding a T-Joint in the Horizontal Welding Position

## Objective:

In this job, you will learn to make a fillet weld on a T-joint in the horizontal welding position.

> **Note**
>
> Do not attempt this job until you have read all safety precautions, satisfactorily completed the *Oxyfuel Gas Cutting (OFC) and Welding (OFW) Safety Test*, and been approved by your instructor.

1. You will need six pieces of low-carbon steel. Four pieces must measure 1/8″ × 1 1/2″ × 6″ (3.2 mm × 40 mm × 150 mm), and two pieces must measure 1/8″ × 3″ × 6″ (3.2 mm × 75 mm × 150 mm). Also obtain two welding rods of the correct size.

2. Fill in the following information for your torch, tip, and metal thickness. Refer to Figure 24-34 in the text.
   A. Correct tip drill size: _____
   B. Tip number size: _____
   C. Oxygen working pressure: _____ psig
   D. Acetylene working pressure: _____ psig
   E. Welding rod diameter: _____
   F. Fillet weld leg size (see drawing for step 5): _____

3. What are the suggested welding angles and distances for this job?
   A. Suggested work angle: _____ (which is an angle of _____ off the surface).
   B. Suggested travel angle: _____ (which is an angle of _____ off the surface).
   C. How far from the base metal is the flame held? _____
   D. At what angle is the welding rod held to the base metal? _____

4. Turn on the welding outfit, set the working pressures, and adjust the flame to neutral.

5. Place two of the metal pieces together to form a T-joint and tack weld these pieces together. Think carefully about the order of assembly, tack welding, and welding before beginning this job. Be sure to make the proper welds according to the welding symbols in the following drawing. Make the required welds in the horizontal position.

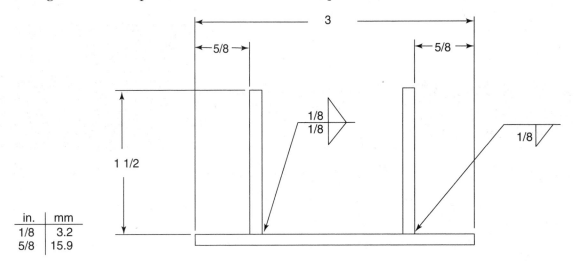

6. After the first two pieces have been welded together, make the second T-joint to complete the weldment. Make the required welds in the horizontal welding position.

7. Repeat steps 5 and 6 on the remaining three pieces of metal. Show these welds to your instructor for a grade.

## Inspection:

Each fillet weld should be inspected before welding the next joint. Each weld bead should be straight and slightly convex, with an even width and evenly spaced ripples. The weld bead should not be undercut or overlapped.

## Assigned Job 28-3
# Welding a Square-Groove Butt Joint in the Horizontal Welding Position

### Objective:

In this job, you will learn to weld a square-groove butt joint in the horizontal welding position. You will use a stringer bead to make this weld.

> **Note**
> Do not attempt this job until you have read all safety precautions, satisfactorily completed the *Oxyfuel Gas Cutting (OFC) and Welding (OFW) Safety Test*, and been approved by your instructor.

1. You will need three pieces of mild steel that measure $1/8'' \times 1\ 1/2'' \times 6''$ (3.2 mm × 40 mm × 150 mm) and two welding rods of the correct diameter.

2. Fill in the following information for your torch, tip, and base metal thickness.
    A. Correct tip drill size: _____
    B. Tip number size: _____
    C. Oxygen working pressure: _____ psig
    D. Acetylene working pressure: _____ psig
    E. Welding rod diameter: _____

3. What are the suggested welding angles and distances for this job?
    A. Suggested work angle: _____ (which is an angle of _____ off the surface).
    B. Suggested travel angle: _____ (which is an angle of _____ off the surface).
    C. How far from the base metal is the flame held? _____
    D. The welding rod is held at what angle to the weld axis? _____

4. Turn on the welding outfit, light the flame, and adjust the flame to neutral.

5.  Place the three pieces on a firebrick. Tack weld each joint in three places to form the weldment shown in the following illustration. However, before tack welding each joint, bend a 1/16″ or 3/32″ (1.6 mm–2.4 mm) filler rod into a U shape and place it into the joint. This maintains the proper root spacing of 1/16″–3/32″ (1.6 mm–2.4 mm), which should be left between the pieces after tack welding is completed. After tack welding the joint, remove the filler rod.

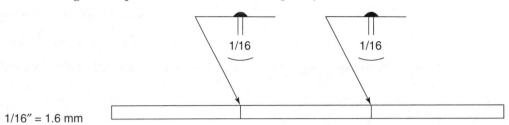

6.  These welds are to be made in the horizontal welding position. Place the tacked weldment into a weld positioner or lean it against a second firebrick or steel block, so that the welds can be made in the correct position.

## Inspection:

Each butt weld should have a small and uniform amount of penetration showing on the side opposite the weld face. Each weld bead should be straight and slightly convex, with an even width, evenly spaced ripples, and no visible inclusions. There should be no sagging of the weld bead at the upper edge and no undercut or overlap.

Name: _____   Date: _____

Class: _____   Instructor: _____

Lesson Grade: _____   Instructor's Initials: _____

# Welding a Lap Joint in the Vertical Welding Position

## Objective:

In this job, you will learn to make a fillet weld on a lap joint in the vertical welding position.

---

**Note**

Do not attempt this job until you have read all safety precautions, satisfactorily completed the *Oxyfuel Gas Cutting (OFC) and Welding (OFW) Safety Test*, and been approved by your instructor.

---

1. Obtain three pieces of low-carbon steel that measure 1/8″ × 1 1/2″ × 6″ (3.2 mm × 40 mm × 150 mm). Also obtain two welding rods of the correct size.

2. Fill in the following information for your torch, tip, and metal thickness.
   A. Correct tip drill size: _____
   B. Tip number size: _____
   C. Oxygen working pressure: _____ psig
   D. Acetylene working pressure: _____ psig
   E. Welding rod diameter: _____
   F. Fillet weld leg size (see drawing for step 5): _____

3. What are the suggested welding angles and distances for this job?
   A. Suggested work angle: _____ (which is an angle of _____ off the surface).
   B. Suggested travel angle: _____ (which is an angle of _____ off the surface).
   C. Describe where the flame should be directed:

   _____

   _____

   D. How far from the weld pool surface should the flame be? _____
   E. At what angle is the welding rod held to the base metal? _____

4. Turn on the welding outfit, set the working pressures, and adjust the flame to neutral.

5. Place the metal on a firebrick. Arrange three pieces of 1/8″ (3.2 mm) metal as shown in the following drawing. Tack weld each lap joint in three places. This may be done in the flat welding position.

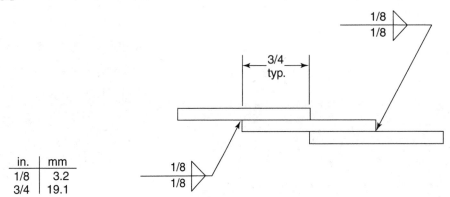

6. Place the weldments in the correct position by clamping them into a positioner. If there is no positioner available, support the weldment with a second firebrick. When making the weld, the weld face and weld axis must be 80°–90° from the straight down position.

7. Make the fillet welds on the 1/8″ (3.2 mm) metal as shown in the drawing.

## Inspection:

Each fillet weld should be straight and slightly convex. It should have an even width and evenly spaced ripples in the weld bead. There should be no evidence of undercut or overlap of the weld bead.

Name: _____  Date: _____

Class: _____  Instructor: _____

Lesson Grade: _____  Instructor's Initials: _____

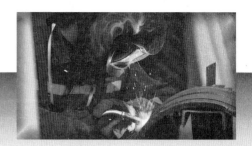

## Assigned Job 28-5
# Welding a T-Joint in the Vertical Welding Position

## Objective:
In this job, you will learn to make a fillet weld on a T-joint in the vertical welding position.

**Note**
Do not attempt this job until you have read all safety precautions, satisfactorily completed the *Oxyfuel Gas Cutting (OFC) and Welding (OFW) Safety Test*, and been approved by your instructor.

1. Obtain four pieces of low-carbon steel that measure 1/16″ × 1 1/2″ × 6″ (1.6 mm × 40 mm × 150 mm) and four pieces that measure 1/8″ × 1 1/2″ × 6″ (3.2 mm × 40 mm × 150 mm). Also obtain two welding rods of the correct size for each metal thickness.

2. Fill in the following information for your torch, tip, and metal thickness.

|  | 1/16″ | 1/8″ |
|---|---|---|
| A. Correct tip drill size: | _____ | _____ |
| B. Tip number size: | _____ | _____ |
| C. Oxygen working pressure: | _____ psig | _____ psig |
| D. Acetylene working pressure: | _____ psig | _____ psig |
| E. Welding rod diameter: | _____ | _____ |
| F. Fillet weld leg size (see drawing in step 5): | _____ | _____ |

3. What are the suggested welding angles and distances for this job?
   A. Suggested work angle: _____ (which is an angle of _____ off the surface).
   B. Suggested travel angle: _____ (which is an angle of _____ off the surface).
   C. How far from the base metal is the flame held? _____
   D. At what angle is the welding rod held to the base metal? _____

4. Turn on the welding outfit, set the working pressures, and adjust the flame to neutral.

5. Place one piece of 1/16″ (1.6 mm) metal on a firebrick, then place the other piece of 1/16″ (1.6 mm) metal on it to form the weldment shown in the drawing A.

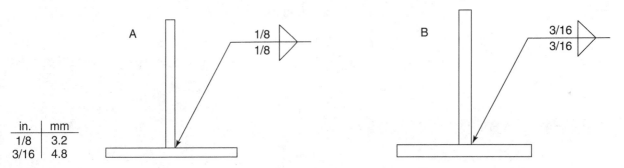

6. Tack weld each side of the joint in three places.

7. Repeat steps 5 and 6 for the 1/8″ (3.2 mm) pieces. Refer to drawing B for step 5.

8. Place the weldment in a positioner, so that each weld can be made in the vertical welding position. If a positioner is not available, support the weldment with a second firebrick so that the weld can be made in the vertical position.

9. Make a fillet weld on both sides of the 1/16″ (1.6 mm) metal as shown by the AWS welding symbols on drawing A in step 5. Make each weld in the vertical welding position.

10. Make a fillet weld on both sides of the 1/8″ (3.2 mm) metal. See drawing B in step 5.

11. Repeat steps 5, 6, 8, and 9, using the remaining pieces of 1/16″ (1.6 mm) metal. This weldment will be used for a grade.

12. Repeat steps 5, 6, 8, and 10, using the remaining pieces of 1/8″ (3.2 mm) metal. This weldment will be used for a grade.

## Inspection:

Each fillet weld should be straight and slightly convex. It should have an even width, and evenly spaced ripples in the weld bead. The weld bead should not be undercut or overlapped.

Name: _____   Date: _____

Class: _____   Instructor: _____

Lesson Grade: _____   Instructor's Initials: _____

# Assigned Job 28-6
# Welding a Square-Groove
# Butt Joint in the Vertical Welding Position

## Objective:

In this job, you will learn to weld a square-groove butt joint in the vertical welding position. You will use a stringer bead to make this weld.

> **Note**
> Do not attempt this job until you have read all safety precautions, satisfactorily completed the *Oxyfuel Gas Cutting (OFC) and Welding (OFW) Safety Test*, and been approved by your instructor.

1. You will need six pieces of mild steel that measure 1/8″ × 1 1/2″ × 6″ (3.2 mm × 40 mm × 150 mm) and two welding rods of the correct diameter.

2. Fill in the following information regarding your torch, tip, and base metal thickness.
   A. Correct tip drill size: _____
   B. Tip number size: _____
   C. Oxygen working pressure: _____ psig
   D. Acetylene working pressure: _____ psig
   E. Welding rod diameter: _____

3. What are the suggested welding angles and distances for this job?
   A. Suggested work angle: _____ (which is an angle of _____ off the surface).
   B. Suggested travel angle: _____ (which is an angle of _____ off the surface).
   C. The welding rod is held at what angle to the weld axis? _____
   D. What is the size of the root opening? See drawing in step 5: _____

4. Turn on the welding outfit, light the flame, and adjust the flame to neutral.

5. Place three 1/8″ (3.2 mm) pieces of steel on a firebrick and tack weld each joint in three places to form the weldment shown in the drawing. Use a U-shaped spacer rod to maintain the specified root opening, which should exist between the pieces after the tack welding is completed.

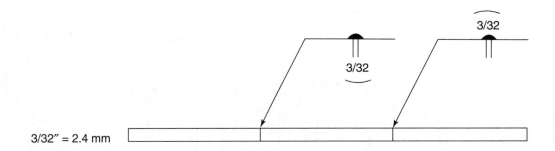

6. Place the tacked weldment in a weld positioner or lean it against a second firebrick so that the welds can be made in the vertical welding position. Make the butt welds as directed by the AWS welding symbols in the drawing.

7. Repeat steps 5 and 6, using the remaining pieces of metal. This weldment will be graded.

## Inspection:

Each butt weld should have a small and uniform amount of penetration showing on the side opposite the weld bead face. Each weld bead is to be straight and slightly convex, with an even width and evenly spaced ripples. There should be no visible inclusions.

Name: _____ Date: _____

Class: _____ Instructor: _____

Lesson Grade: _____ Instructor's Initials: _____

## Assigned Job 28-7
# Welding a Lap Joint in the Overhead Welding Position

## Objective:
In this job, you will learn to make a fillet weld on a lap joint in the overhead welding position.

**Note**
Do not attempt this job until you have read all safety precautions, satisfactorily completed the *Oxyfuel Gas Cutting (OFC) and Welding (OFW) Safety Test*, and been approved by your instructor.

1. Obtain five pieces of low-carbon steel that measure 1/8″ × 1 1/2″ × 6″ (3.2 mm × 40 mm × 150 mm). Also, obtain three welding rods of the correct size.

2. Fill in the following information for your torch, tip, and metal thickness.
   A. Correct tip drill size: _____
   B. Tip number size: _____
   C. Oxygen working pressure: _____ psig
   D. Acetylene working pressure: _____ psig
   E. Welding rod diameter: _____
   F. Fillet weld leg sizes (see drawing in step 5):

   _____

3. What are the suggested welding angles and distances for this job?
   A. Suggested work angle: _____ (which is an angle of _____ off the surface).
   B. Suggested travel angle: _____ (which is an angle of _____ off the surface).
   C. Describe where the flame should be directed.

   _____

   _____

   D. How far from the weld pool surface should the flame be? _____
   E. At what angle is the welding rod held to the base metal? _____

4. Turn on the welding outfit, set the working pressures, and adjust the flame to neutral.

5. Place two pieces of metal on a firebrick. Arrange pieces A and B as shown in the drawing in the following figure. Tack weld each lap joint in three places. The tack welding may be done in the flat welding position.

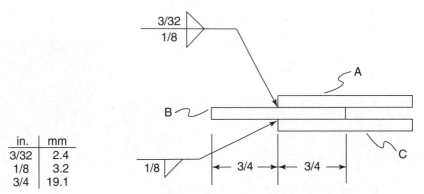

| in. | mm |
|-----|------|
| 3/32 | 2.4 |
| 1/8 | 3.2 |
| 3/4 | 19.1 |

6. When positioning the weld, the weld face and weld axis must be within 5° of horizontal. Hold the weldment in the correct position by clamping it in a welding positioner.

7. Make the fillet welds on parts A and B as shown in the drawing. Turn the weldment as needed to make both welds in the overhead welding position. Keep the weld pool small to prevent sagging and to prevent overlap of the weld bead onto the surface of piece B.

8. These two welds are intended for practice.

9. Repeat steps 5, 6, and 7 using the remaining pieces. This weldment will be graded. It will be combined with a third piece to complete the weldment shown in step 5.

10. Place piece C on piece B and tack weld the joint in three places. The tack welding may be done in the flat welding position.

11. Weld piece C and B as shown in step 5. There can be no overlap in the first welds or the surface of piece C will not fit tightly on surface B.

## Inspection:

Each fillet weld should be straight and slightly convex. It should have an even width and evenly spaced ripples in the weld bead. There should be no evidence of undercut or overlap of the weld bead.

Name: _____   Date: _____

Class: _____   Instructor: _____

Lesson Grade: _____   Instructor's Initials: _____

## Assigned Job 28-8
# Welding a T-Joint in the Overhead Welding Position

## Objective:

In this job, you will learn to make a fillet weld on a T-joint in the overhead welding position.

> **Note**
> Do not attempt this job until you have read all safety precautions, satisfactorily completed the *Oxyfuel Gas Cutting (OFC) and Welding (OFW) Safety Test*, and been approved by your instructor.

1. Obtain six pieces of low-carbon steel that measure 1/16″ × 1 1/2″ × 6″ (1.6 mm × 40 mm × 150 mm). Also obtain two welding rods of the correct size.

2. Fill in the following information for your torch, tip, and metal thickness.
   A. Correct tip drill size: _____
   B. Tip number size: _____
   C. Oxygen working pressure: _____ psig
   D. Acetylene working pressure: _____ psig
   E. Welding rod diameter: _____
   F. Fillet weld leg sizes (see drawing in step 5): _____

3. What are the suggested welding angles and distances for this job?
   A. Suggested work angle: _____ (which is an angle of _____ off the surface).
   B. Suggested travel angle: _____ (which is an angle of _____ off the surface).
   C. How far from the base metal is the flame held? _____
   D. At what angle is the welding rod held to the base metal? _____

4. Turn on the welding outfit, set the working pressures, and adjust the flame to neutral.

5. Place three pieces of the metal on a firebrick. Tack weld them to form the weldment shown in the drawing.

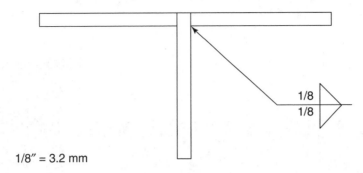

1/8" = 3.2 mm

6. Place the weldment in a weld positioner, so that each weld is made in the overhead welding position.

7. Make a fillet weld on both sides of the metal, as shown by the AWS welding symbols on the drawing in step 5. These welds will be done for practice. Turn the weldment as needed to make both welds in the overhead welding position.

8. Repeat steps 5, 6, and 7 with the remaining three pieces of metal. These welds will be graded.

## Inspection:

Each fillet weld should be straight and slightly convex, with an even width and evenly spaced ripples in the weld bead. The weld bead should not be undercut or overlapped.

## Assigned Job 28-9

# Welding a Square-Groove and V-Groove Butt Joint in the Overhead Welding Position

### Objectives:

In this job, you will learn to weld a square-groove butt joint and a V-groove butt joint on an outside corner. Both welds will be made in the overhead welding position. You will use a stringer bead to make these welds.

> **Note**
>
> Do not attempt this job until you have read all safety precautions, satisfactorily completed the *Oxyfuel Gas Cutting (OFC) and Welding (OFW) Safety Test*, and been approved by your instructor.

1. You will need six pieces of mild steel that measure 1/8″ × 1 1/2″ × 6″ (3.2 mm × 40 mm × 150 mm) and two welding rods of the correct diameter.

2. Fill in the following information regarding your torch, tip, and base metal thickness.
   A. Correct tip drill size: _____
   B. Tip number size: _____
   C. Oxygen working pressure: _____ psig
   D. Acetylene working pressure: _____ psig
   E. Welding rod diameter: _____

3. What are the suggested welding angles and distances for this job?
   A. Suggested work angle: _____ (which is an angle of _____ off the surface).
   B. Suggested travel angle: _____ (which is an angle of _____ off the surface).
   C. At what angle to the weld axis is the welding rod held? _____
   D. What is the size of the root opening for each joint? See the drawing in step 5. _____

4. Turn on the welding outfit, light the flame, and adjust the flame to neutral.

5.  Place three pieces of metal on a firebrick. Tack weld each joint in three places to form the weldment shown in the drawing. U-shaped metal spacers may be used to control the root opening of the square-groove butt joint while tack welding. The correct root opening must exist after the tack welding has been completed.

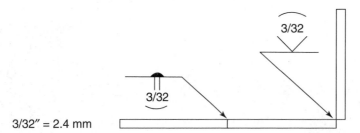

3/32" = 2.4 mm

6.  Place the tacked weldment in a weld positioner and adjust as needed so that each weld can be made in the overhead welding position.

7.  Weld each joint as required by the AWS welding symbol shown in the drawing. Keep the weld pool small to prevent sagging. The V-groove weld may require more than one pass to fill the joint. This weldment is made for practice.

8.  Repeat steps 5, 6, and 7 with the remaining three pieces of metal for a grade.

## Inspection:

Each butt weld should have a small and uniform amount of penetration showing on the root side. Each weld bead is to be straight and slightly convex, with an even width and evenly spaced ripples. There should be no visible inclusions. The weld bead on the outside corner joint must fill the V-groove from edge to edge.

## Lesson 29
# Brazing and Braze Welding

## Objectives:

You will be able to select the correct brazing filler metal and brazing flux for brazing various metals. You will also be able to make an acceptable brazed or braze-welded joint.

## Instructions:

Read Chapter 29 and study Figures 29-1 through 29-12. Then, answer or complete the following questions.

1. *True or False?* The temperature at which the filler metal begins to melt is called the liquidus temperature.

_____   2. A thick layer of brazing filler metal is added when ____.
A. brazing
B. braze welding

_____   3. For brazing or braze welding, the filler metal must melt at a temperature ____.
A. above 840°F (450°C)
B. below 840°F (450°C)
C. above 1544°F (840°C)
D. below 1544°F (840°C)

4. Brazing filler metal comes in a variety of shapes or forms. Name three of these forms.

_____

_____

_____

5. Name five gases that have been used successfully as the fuel gas in oxyfuel gas brazing.

_____

_____

*For Questions 6–11, complete the statements regarding the recommended angles and distances for braze welding.*

6. The recommended work angle for braze welding a butt joint in the flat position is _____, which can also be described as an angle of _____ off the surface.

7. The recommended travel angle for braze welding a butt joint in the flat position is _____, which can also be described as an angle of _____ off the surface.

8. The recommended work angle for braze welding an inside corner joint in the flat position is _____, which can also be described as an angle of _____ off the surface.

9. The recommended travel angle for braze welding an inside corner joint in the flat position is _____, which can also be described as an angle of _____ off the surface.

10. At what angle is the brazing rod held to the base metal? _____

11. How far from the base metal should the flame tip be held? _____

12. *True or False?* Silver and gold alloys are used as brazing filler metals.

_____ 13. Which of the following safety considerations for brazing or braze welding is not true?
   A. Many fluxes are harmful to the skin.
   B. When acids are required for cleaning, special training is required.
   C. Cadmium fumes have a strong garlic odor.
   D. Cadmium oxide fumes are very dangerous if inhaled. They may even cause death.
   E. The exposure limit for cadmium fumes is 0.1 milligrams per cubic meter of air for a daily eight-hour period.

14. *True or False?* In braze welding, the width of the weld bead is controlled by the flame height only.

15. What is the definition for a flux?

_____

_____

_____

16. In the brazing process, ____ ____ draws the filler metal into the space between the parts of the brazement to form a bond.

_____

17. For brazing 1/8″ (3.2 mm) metal, the correct suggested torch tip sizes are #____ or #____.

_____ 18. Flux may be added to the base metal or filler rod. Which of the following is not an acceptable method of applying flux?
   A. Applying it to the surface with a brush.
   B. Dipping the base metal into the flux.
   C. Applying it to the filler rod when the filler rod is manufactured.
   D. Applying it to the filler rod during the brazing/braze welding process.

19. What type of flame is recommended for brazing or braze welding?

_____

20. *True or False?* When brazing, a large area of the base metal around the joint is heated before the filler metal is added.

21. Name four ways of mechanically cleaning a joint area before brazing or braze welding.

_____

_____

Name _____

_____ 22. For a given thickness of base metal, which process requires the larger torch tip orifice?
    A. Brazing.
    B. Braze welding.

_____ 23. Which of the following does not apply to brazing and braze welding?
    A. It is done at temperatures far below the temperatures required for welding.
    B. The strength of brazed and braze-welded parts is less than that of fusion-welded parts.
    C. Metals may be joined without changing their heat-treatment characteristics.
    D. Brazing assemblies are called weldments.
    E. This process may be used more easily to join thin metal sections.

24. *True or False?* Brazing filler rods come in at least six diameters from 1/16″ to 1/4″ (1.6 mm and 6.4 mm); they are normally 36″ (.91 m) long.

_____ 25. Which of the following is an acceptable range of clearances between the parts in a brazement?
    A. .001″–.005″ (.025–.127 mm).
    B. .002″–.010″ (.051–.254 mm).
    C. .003″–.005″ (.076–.127 mm).
    D. .005″–.010″ (.127–.254 mm).
    E. Any of the above.

## Assigned Job 29-1
# Brazing a Lap Joint on Thin Mild Steel in the Flat Position

## Objectives:

In this job, you will learn to make an acceptable brazed lap joint on thin mild steel.

> **Note**
> Do not attempt this job until you have read all safety precautions, satisfactorily completed the *Oxyfuel Gas Cutting (OFC) and Welding (OFW) Safety Test*, and been approved by your instructor.

1. Obtain six pieces of mild steel that measure 1/16″ × 1 1/2″ × 6″ (1.6 mm × 40 mm × 150 mm).

2. Fill in the following information for the job to be done: Refer to Figure 29-7.
   A. Suggested tip size for brazing: _____
   B. Recommended oxygen pressure: _____ psig
   C. Recommended acetylene pressure: _____ psig
   D. Recommended brazing rod composition: _____
   E. Recommended filler rod diameter: _____
   F. Four suggested types of flux:

   _____

   _____

   G. How far from the base metal should the flame tip be held for brazing? _____

3. Obtain a brazing filler rod of the correct diameter and a container of the correct flux.

4. Thoroughly clean both sides of the six pieces of base metal using any available method suggested in the Preparing to Braze or Braze Weld section of the text.

5. Place one clean piece of metal on another so that they overlap 3/4″ (19 mm). Then, clamp them in place with small C-clamps or locking pliers. The pieces must be flat and fit tightly together.

6. Turn on your oxyacetylene welding outfit, set the correct pressures, and light the torch. Adjust the flame to neutral or slightly carburizing.

7. The flux may be added to the joint in a variety of ways. The following method is the suggested method for this job:
   A. Place a clean piece of metal on your welding table.
   B. Sprinkle the correct flux on this metal until the flux layer is about 1" (25 mm) wide and 1/8" (3.0 mm) deep.
   C. Heat the end of your filler rod for a distance of about 2" (50 mm). Then, roll the rod in the flux on your clean metal plate. You will notice that the flux adheres to the part of the rod that you heated.
   D. The filler rod must always have flux on it. Whenever the rod needs more flux, reheat it and roll it in the flux on your clean metal plate.

8. Start brazing at the right-hand edge of the lap joint if you are right-handed or at the left-hand edge if you are left-handed. Point the torch flame at the lap joint and heat both pieces of metal to a bright red. Touch the filler rod to the joint. If the filler rod melts, the metal is at the correct brazing temperature. If it does not melt, withdraw the filler rod and continue to heat the base metal. Then, retest the temperature with the filler rod. Once the correct temperature is reached, continue across the joint slowly, heating and adding filler metal until the entire joint has been completed.

9. If the brazing material spreads over too wide an area, withdraw the flame and let the metal cool for a few seconds. Then, reheat the area and begin to braze again.

10. Braze on only one side of the joint.

11. Complete three brazed lap joints with your six pieces of metal.

## Inspection:

The brazed joint should have a thin layer of brazing material on the surface of both pieces. Filler material should also be seen between the upper and lower surfaces. The width of the filler metal should be uniform, with no overheating showing. The color of the completed joint should be the same color as the filler rod.

## Assigned Job 29-2
# Braze Welding an Outside Corner Joint in the Flat Welding Position

## Objective:

In this job, you will learn to make an acceptable braze-welded outside corner joint in the flat welding position.

> **Note**
> Do not attempt this job until you have read all safety precautions, satisfactorily completed the *Oxyfuel Gas Cutting (OFC) and Welding (OFW) Safety Test*, and been approved by your instructor.

1. Obtain four pieces of 3/16″ (4.8 mm) mild steel that measure 1 1/2″ × 6″ (40 mm × 150 mm).

2. Clean one edge on each piece of steel. Clean 3/4″ (19 mm) of the upper and lower surfaces along these clean edges using one of the mechanical or chemical cleaning processes described in the text.

3. Fill in the following information for the metal being used:
   A. Recommended torch tip size: _____
   B. Recommended oxygen pressure: _____ psig
   C. Recommended acetylene pressure: _____ psig
   D. Recommended composition of brazing rod:

   _____

   E. Recommended filler rod diameter: _____
   F. List four recommended fluxes:

   _____

   _____

   _____

   _____

4. Obtain one uncoated brazing filler rod and one flux-coated brazing filler rod of the correct diameter, and a container of the correct braze-welding flux.

5. Turn on the oxyacetylene welding outfit, set the correct pressure, and adjust to a neutral or slightly carburizing flame.

6. Arrange two cleaned pieces to make an outside corner joint in the flat welding position, as shown in the following drawing.

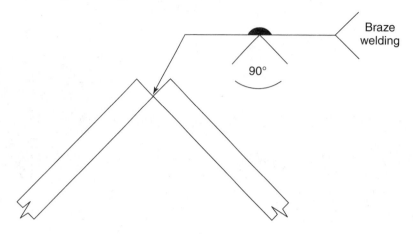

7. Add the flux to the uncoated filler rod using the following technique:
   A. Place a clean piece of metal on the welding table.
   B. Sprinkle the correct flux on this metal until the flux layer is about 1″ (25 mm) wide and 1/8″ (3.0 mm) deep.
   C. Heat the end of your filler rod for a distance of about 2″ (50 mm). Then, roll the end of the rod in the flux on the clean metal plate. You will notice that the flux adheres to the part of the rod that you heated.
   D. The filler rod must always have flux on it. Whenever the rod needs more flux, reheat it and roll it in the flux on the clean metal plate.

8. Tack the brazement, using the uncoated filler rod.

9. Start the braze weld at the right or left edge (depending on whether you are right-handed or left-handed). Heat the base metal to a bright red and test the temperature with the filler rod. When the rod melts, the base metal is at the correct temperature for braze welding. Add the filler rod as often as required to create a convex bead that covers the joint from edge to edge. Move the torch along the joint and continue to add the filler rod until the braze weld is completed. If the bead becomes too wide, withdraw the flame and allow the metal to cool for a few seconds. Then, continue the bead.

**Note**
To create a full strength braze-welded joint, 100% penetration must be obtained.

10. Use the flux-coated filler rod to complete a second brazement like the one shown in step 6.

## Inspection:

The braze-welded bead must be straight and have a uniform width. It must have evenly spaced ripples and 100% penetration. The completed braze-welded bead should be the same color as the filler rod. If the bead is copper colored, the bead was overheated.

Name: _____  Date: _____

Class: _____  Instructor: _____

Lesson Grade: _____  Instructor's Initials: _____

# Lesson 30
# Soldering

## Objectives:

You will be able to describe how to select the solder and flux for a given soldering job. You will be able to solder a copper pipe joint. You will become familiar with the various torches and torch tips used in soldering.

## Instructions:

Read Chapter 30 and study Figures 30-1 through 30-11. Then, answer or complete the following questions.

1. Solder is drawn into the small spaces between the heated metal by _____ action.

   _____

2. What type of flux is recommended when soldering a metal with an applied finish of cadmium?

   _____

_____ 3. Solder is generally applied in thin layers. Which of the following is a reason for applying solder in thick layers?
   A.  Improved adhesion.
   B.  To improve the looks of the part.
   C.  For strength.
   D.  All of the above.

_____ 4. Which type of flux is the most corrosive and must not be used on electrical or electronic parts?
   A.  Organic.
   B.  Inorganic.
   C.  Rosin.
   D.  None of the above.
   E.  All of the above.

5. List four fuel gases that may be used with an air-fuel gas torch when soldering.

   _____

   _____

   _____

   _____

_____     6.   What number filter lens is suggested for light soldering operations?
           A.   #2–#4.
           B.   #3–#5.
           C.   #1–#6.
           D.   #10–#12.
           E.   #8.

7.  Soldering differs from brazing because of the temperature at which it is done. Soldering is done at a temperature below ____°F (____°C).

_____

_____     8.   Which completed joint is the strongest?
           A.   A soldered joint.
           B.   A brazed joint.
           C.   A welded joint.
           D.   A braze-welded joint.

9.  *True or False?* The liquidus temperature is the temperature at which the soldering filler metal is completely melted.

10. What are the two purposes of a flux?

_____

_____

11. *True or False?* Public Law 99-339 effectively bans the use of lead solder, lead-bearing fluxes, and lead pipes from use in all drinking water systems.

12. The three classifications of flux are ____, ____, and ____.

_____

_____     13.  What type of solder flows into small spaces most easily?
           A.   One that has the solidus temperature and liquidus temperatures far apart.
           B.   One that has the solidus and liquidus temperatures close together.
           C.   One with the highest liquidus temperature.
           D.   One with the highest solidus temperature.

14. *True or False?* Rosin-based fluxes are the most effective cleaners.

15. Which type of flux is recommended for soldering aluminum?
           A.   Organic.
           B.   Inorganic.
           C.   Rosin.
           D.   Special.

16. List five methods used to clean the base metal surface before soldering.

_____

_____

_____

17. Soldering requires a wide flame to evenly heat the base metal. When using an oxyfuel gas outfit to solder, a torch tip should be chosen with a tip orifice between ____ and ____.

_____

Name _____

_____ 18. New lead-free solders have a ____ difference between their solidus and liquidus.
   A.  10°F–30°F (6°C–17°C)
   B.  20°F–40°F (11°C–22°C)
   C.  30°F–50°F (17°C–28°C)
   D.  40°F–60°F (22°C–33°C)
   E.  50°F–70°F (28°C–39°C)

_____ 19. The clearance between parts in a soldered joint is commonly ____.
   A.  .025″–.127″ (.64 mm–3.2 mm)
   B.  .001″–.003″ (.03 mm–.08 mm)
   C.  .003″–.005″ (.08 mm–.13 mm)
   D.  .635″–.750″ (16 mm–19 mm)
   E.  .005″–.010″ (.13 mm–.25 mm)

20. *True or False?* Lead-based solders are required for use in drinking water systems.

21. *True or False?* The soldering process requires high temperatures in order to partially melt the base metal. This creates a strong molecular bond with the melted solder.

22. *True or False?* During soldering, the torch flame should be directed at both the base metal and the solder in order to melt both metals.

23. *True or False?* Soldered joints have excellent electrical and thermal conductivity.

24. *True or False?* Joints to be soldered should be cleaned chemically (with flux) but not mechanically (with wire brush, sandpaper, or abrasive blasting) because mechanical cleaning removes base metal and causes the joint to fit loosely.

*The standard steps in the soldering process are listed below, but are not in the correct order. For Questions 25–31, place each step in the correct order by writing the appropriate number (1–7) in the space to the left of the step.*

_____ 25. Apply the appropriate flux to the surfaces to be joined.

_____ 26. Light the torch and apply heat to the joint.

_____ 27. Check the fit of the parts to be joined. Be sure they have close contact with each other.

_____ 28. Properly clean the metals. All oxides, grease, paint, and dirt must be removed.

_____ 29. Add solder to the heated joint and work as quickly as possible.

_____ 30. Remove excess flux from the joint to prevent corrosion.

_____ 31. Be sure the metals to be soldered are firmly supported.

## Assigned Job 30-1
# Soldering Copper Pipe Fittings

## Objective:

In this job, you will learn to solder 3/4" (19 mm) diameter copper pipe fittings using lead-free solder.

**Note**

Do not attempt this job until you have read all safety precautions, satisfactorily completed the *Oxyfuel Gas Cutting (OFC) and Welding (OFW) Safety Test*, and been approved by your instructor.

1. Obtain the following items:
   - Three pieces of 3/4" (19 mm) copper pipe that are 3" (75 mm) long.
   - A 3/4" (19 mm) coupling, 90° elbow, an end cap, and a threaded pipe union.
   - A roll of 1/8" (3.2 mm), 95.5% tin, 4% copper, .05% silver or other lead-free solder.
   - The correct flux and a flux brush.
   - A piece of sandpaper or abrasive cloth.

2. Clean the ends of the pipe for a distance of about 1" (25 mm). Clean the inside of the fittings also.

3. Apply the correct flux to the ends of the pipes and assemble the fittings as shown in the drawing.

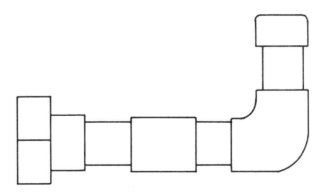

4. Fill in the following information for the soldering job being done with an oxyacetylene flame. Figures 24-34 and 30-6 provide needed reference information.
   A. Recommended type of flux (same as for lead-tin solder): _____
   B. Recommended tip orifice size (assuming the pipe thickness is 1/16" (1.6 mm): _____
   C. Oxygen pressure: _____ psig
   D. Acetylene pressure: _____ psig
   E. How far away from the joint should the flame be held? _____
   F. What type of flame should be used? _____

5. Solder the fittings to the pipes as shown in drawing for step 3. Support the parts while they are soldered and until they are thoroughly cooled.

6. Use a clean cloth to wipe all excess solder from each joint while the solder is still hot.

**Caution**
Be careful not to burn your hands while wiping away the excess solder.

7. After the soldered assembly has cooled, attach the threaded pipe union to a cold water line. Turn on the water and check the assembly for leaks.

## Inspection:

With the water turned on, the pipe fittings should show no leakage. Excess solder should not be visible around the pipes or fittings.

Name: _____  Date: _____

Class: _____  Instructor: _____

Lesson Grade: _____  Instructor's Initials: _____

# Section 7
# Resistance Welding

## Resistance Welding Safety Test

### Objectives:

You will be able to discuss the potential safety hazards of resistance welding. You will also be able to describe the safety precautions required when working with resistance welding equipment.

### Instructions:

Always follow safe practices when welding. If you have safety questions or concerns, ask your instructor. This test does not include questions about every safety topic, but is intended to highlight key items. Review Chapters 31 and 32 and Figures 31-1 through 31-13 and 32-1 through 32-10. Pay special attention to the Safety section in Chapter 31. Then, answer or complete the following questions.

1. *True or False?* The forces used in resistance welding can cause bodily harm.

2. What part of the welding machine applies the force to the workpiece?

   _____

3. To prevent injury when using a resistance welding machine, where should you not put any part of your body?

   _____

4. *True or False?* The currents and voltages used in resistance welding are high enough to cause an electrical shock.

_____   5. How high can current be during a resistance weld?
   A. A few amps.
   B. A few hundred amps.
   C. A few thousand amps.
   D. Tens of thousands of amps.

6. Four different times may be used during a spot weld sequence. During which of these times does electrical current flow?

   _____

7. What is a weld called when molten metal squirts out from it?

   _____

8. What types of protective clothing should you wear when resistance welding or when working near resistance welding equipment?

_____

_____

9. What should you do if an air or a hydraulic leak develops on the resistance welding equipment you are using?

_____

10. Describe four safety checks that should be performed regularly when you are using a resistance welding machine.

_____

_____

_____

_____

## Task:

Your instructor will demonstrate how to turn the welding machine on and off. Your instructor may also demonstrate how to turn the circuit breaker on and off. You may be asked to perform these tasks yourself.

# Lesson 31
# Resistance Welding:
# Equipment and Supplies

## Objectives:

You will be able to describe the operation of resistance welding equipment and discuss selection of the proper electrodes for use in resistance welding.

## Instructions:

Read Chapter 31 and study Figures 31-1 through 31-13. Then, answer or complete the following questions.

_____ 1. Where is most of the heat developed during resistance spot welding?
   A. In the top piece of metal.
   B. In the middle of each piece.
   C. At the surface where the two pieces touch.
   D. The heat is equally distributed throughout the base metal.

_____ 2. Which of the following is formed during resistance spot welding?
   A. A weld nugget.
   B. A fillet weld.
   C. A bevel-groove butt weld.
   D. An edge weld.

3. List the three different time intervals set on a resistance welding controller.

_____

_____

_____

_____ 4. Which of the following is not a common design for a resistance welding machine?
   A. Rocker arm.
   B. Gantry.
   C. Press.
   D. Portable Gun.

5. Label the parts of this resistance welding machine.

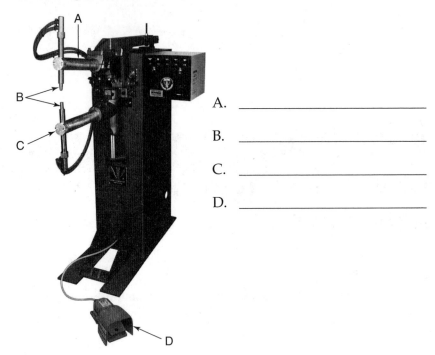

A. _____

B. _____

C. _____

D. _____

6. *True or False?* Large changes in welding current may be obtained by changing the tap setting on the transformer.

7. The percent heat control controls the_____ within one tap setting.

_____

*For Questions 8–14, match the electrode class at right with each condition listed at left. Answers may be used more than once.*

_____    8.   Conductivity of 31%, hardness of 90B.

_____    9.   Most widely used electrode material.

_____   10.   Has the highest electrical conductivity.

_____   11.   Has the highest thermal conductivity.

_____   12.   Molybdenum electrodes with a conductivity of 31% and hardness of 90B.

_____   13.   Very hard, made of tungsten.

_____   14.   Have electrical conductivity of 48% and hardness of 100B.

A.   Class 1.

B.   Class 2.

C.   Class 3.

D.   Class 11.

E.   Class 13.

F.   Class 14.

15. *True or False?* Safety glasses must be worn when resistance welding.

Name: _____  Date: _____

Class: _____  Instructor: _____

Lesson Grade: _____  Instructor's Initials: _____

## Assigned Job 31-1
# Identifying the Parts of a Resistance Welding Machine

## Objectives:

In this job, you will learn the parts of a resistance welding machine. You will also learn about the different switches on the machine.

**Note**

Do not attempt this job until you have read all safety precautions, satisfactorily completed the *Resistance Welding Safety Test*, and been approved by your instructor.

## Instructions:

Your instructor will specify equipment to use for this job. You will need to examine the equipment in order to answer the following questions. Make sure the machine is off and that the circuit breaker is off. To answer the first two questions, look at the machine's nameplate.

1. Who is the manufacturer of the welding machine?

   _____

2. What supply or input voltage does the machine require? _____ volts

3. What type of machine are you studying: rocker, press, or gun?

   _____

4. What method is used to apply the force: air pressure, servo motor or another method?

   _____

5. Locate a regulator, electric controls, or another way to adjust force. How is the force adjusted?

   _____

6. Locate the tap switch. How many tap settings are on the tap switch? _____

7. What company manufactured the controller on the machine?

   _____

8. Does the controller have the ability to set times for the following?
   A. Squeeze      Yes      No
   B. Weld         Yes      No
   C. Hold         Yes      No
   D. Off          Yes      No

9. Does the controller have a way to change the percent heat? Yes or No?

10. What type of electrodes are used: one-piece or caps and adaptors?

_____

11. Does the machine use water for cooling? Yes or No? _____ If "Yes," locate the valve used to turn the water on and off.

## Inspection:

You should be familiar with the machine you have been working on. Your instructor may ask you to point out different parts of the machine and ask you questions about the machine. Be ready to answer them.

Name: _____  Date: _____

Class: _____  Instructor: _____

Lesson Grade: _____  Instructor's Initials: _____

# Lesson 32
# Resistance Welding: Procedures

## Objectives:

You will be able to set up and adjust resistance welding equipment. You also will be able to describe the correct settings for making a good resistance weld.

## Instructions:

Read Chapter 32 and study Figures 32-1 through 32-10. Then, answer or complete the following questions.

1. Figure 32-1 is useful as a starting point to make resistance spot welds. Which Class of welds, A, B, or C, has the largest fusion zone and the highest tension shear strength?

   _____

_____ 2. Which of the following statements is the most correct?
   A. When welding thicker metals, more current is required.
   B. When welding thinner metals, less force but more current is required.
   C. When welding thicker metals, more force and current are required.
   D. When welding thicker metals, less force and current are required.

_____ 3. Electrode tips must be cleaned regularly. Which of the following is a reason to clean the electrode tips?
   A. To remove dirt and oxides.
   B. To keep the tip the correct diameter (prevent flaring of the tip).
   C. To remove metal particles from the tip.
   D. All of the above.
   E. None of the above.

4. *True or False?* The diameter of a completed spot weld is slightly larger than the diameter of the electrode.

5. The force applied by the electrodes through an air cylinder or a hydraulic cylinder is related to the pressure going to the cylinder. The higher the pressure to the cylinder, the _____ the force at the electrodes.

   _____

6. Typical settings for hold time duration are _____ cycles, and typical settings for squeeze time duration are _____ cycles.

   _____

_____    7.  The setup for welding aluminum is different than for welding steel. When aluminum
              is to be welded, _____ are required.
              A.  more current and more weld time
              B.  more current and less weld time
              C.  less current and less weld time
              D.  less current and more weld time

_For Questions 8–11, match the correct action at the right with the situation at left. The situations occur when
using the trial-and-error method for determining the correct weld current:_

_____    8.  You make an expulsion weld.                    A.  Turn up the tap setting.

_____    9.  The parts weld, but the weld size is small.    B.  Turn down the tap setting.
                                                              C.  Turn up the percent heat.
_____   10.  The parts weld and the weld size is slightly
               smaller than the diameter of electrode.        D.  Turn down the percent heat.

                                                              E.  The correct current is set. No
_____   11.  The parts do not weld at all.                      action is needed.

12.  Describe projection welding.

_____

_____

13.  List two industries that use a lot of spot welding.

_____

_____

## Assigned Job 32-1
# Adjusting the Resistance Welding Machine

## Objective:

In this job, you will learn to adjust and set the different parts of a resistance welding machine.

> **Note**
>
> Do not attempt this job until you have read all safety precautions, satisfactorily completed the *Resistance Welding Safety Test*, and been approved by your instructor.
>
> Your instructor will specify a piece of equipment to use for this job and instruct you on the safety features of the equipment. After passing the *Resistance Welding Safety Test* and obtaining permission from your instructor, you may turn on the welding machine. Place the weld/no weld switch in the no weld position.

1. Before continuing the exercise, answer the following questions:

    A. Who is the manufacturer of the welding machine? _____

    B. What type of welding machine will you be studying? _____

    C. What method is used to apply the force? _____

2. Your instructor will demonstrate how the force is set on your machine. After learning how to do so, use a force gauge to set a force of 50 pounds (220 N).

3. Next, readjust the force to 125 pounds (560 N).

4. Adjust the force once again to 400 pounds (1780 N).

5. Locate the tap switch. How many settings are on the tap switch? _____

6. Turn the tap switch to its highest position, then turn it to its lowest position. Make sure the switch is set on a specific position, not between two positions.

7. Locate the way to turn the cooling water on and off. Turn the water on. You may be able to hear it flow. Remember, a machine that is equipped with water cooling must have the water on to prevent overheating.

## Inspection:

You should be able to turn on the welding machine and adjust the force to a given setting. You should be able to adjust the tap switch and be able to turn on the cooling water. Each of these must be done every time you turn on the welding machine. Your instructor may ask you to show how to adjust the equipment you have been working on. Be ready to do so.

## Assigned Job 32-2
# Making Resistance Spot Welds in Mild Steel

## Objective:

In this job, you will learn to resistance spot weld mild steel.

**Note**

Do not attempt this job until you have read all safety precautions, satisfactorily completed the *Resistance Welding Safety Test*, and been approved by your instructor.

1. Obtain about 20 pieces of low-carbon steel that measure 1/16″ × 1″ × 3″ (1.6 mm × 25.4 mm × 75 mm). Another thickness or different material may be specified by your instructor.

2. Clean the material to remove any rust, dirt, or grease. This can be done using steel wool, a wire brush, or a degreasing solution.

3. Before setting up the welding machine, refer to Figure 32-1 in the text and answer the following questions about making a class A weld.

   A. What electrode diameter ("D" in Figure 32-1) should be used? _____

   B. What tip face diameter ("d" in Figure 32-1) should be used? _____

   C. How much force should be used? _____

   D. How many cycles of weld time should be used? _____

   E. How much current should be used? _____

4. Select a pair of electrodes and install them in the machine. Make sure they are installed properly.

5. Use a force gauge. Adjust the air or hydraulic pressure, or the force spring, to obtain the correct force, determined in step 3C.

6. Next, set up a welding schedule. Remember that a complete schedule has four times. Squeeze time must be long enough for the electrodes to close on the metal being welded. Use about 30 cycles for squeeze time. Set the hold time for 30 cycles, as well. Set the off time to zero. Set the weld time to the amount determined in step 3D.

7. If the welding machine is water-cooled, turn on the cooling water.

8. Setting the proper current is an important step. Reread the Making Current Settings section in Chapter 32 of the text. If you have equipment that allows you to set the proper current, use it to adjust the welding machine to the current recorded in step 3E. If you do not have such equipment, use the following steps to set the current.

   A. Set the tap switch on low. If the machine has a number of tap settings, set the switch on a low value (such as 2 or 3).

B. Set the percent heat to 70%.

C. Lay one piece of metal to be welded on top of the other. Each piece should extend 1/2″ (13 mm) beyond the other, as shown in the drawing.

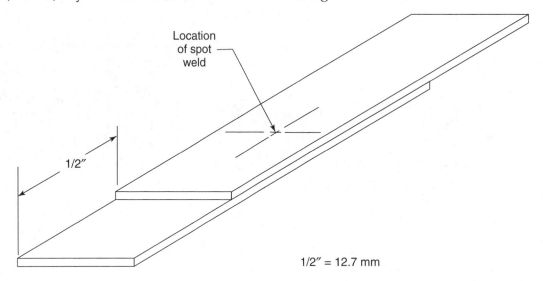

Location of spot weld

1/2″

1/2″ = 12.7 mm

D. Make one spot weld in the middle of the overlapped area.

E. If molten metal squirts out, either the current is too high or the force is too low. If the force is set correctly, turn the tap switch down. If the two pieces of metal do not weld together at all, turn the tap switch up.

F. Weld another weld and make additional changes to the tap switch setting as needed.

G. Look at the spot weld surface. The electrodes should indent the metal slightly. The thickness of the metal where the spot weld was made should be 70%–85% as thick as the base metal. If the electrodes do not indent the metal at all, increase the force. If the electrodes are squeezing the metal to less than 70% of its original thickness, reduce the force.

H. Once the tap switch is set correctly, further adjustments are made with the percent heat setting, not the tap switch. If the two pieces are welded together, tear them apart and look at the size of the weld. It should be slightly smaller than the diameter of the electrode. When welding two pieces of 1/16″ (1.6 mm) material, the diameter of the spot weld should be 3/16″–1/4″ (4.8 mm–6.4 mm). If the weld is too small, turn up the percent heat setting. If the weld is almost the same size as the electrode tip, turn the percent heat down.

**Caution**

When tearing a weld apart, use a protective screen to protect other workers from flying metal.

9. The welding machine is now set to weld. Weld six pairs of metal pieces. Show your work to your instructor.

10. Tear each pair apart. Are all six welds about the same size? _____

11. Write down the tap switch setting, percent heat, and the force you used.

   A. Tap setting: _____

   B. Percent heat: _____

   C. Force: _____

## Inspection:

Show your instructor the pieces after they are torn apart. Your instructor will look at the quality of the welds and give you a grade.

Name: _____     Date: _____

Class: _____     Instructor: _____

Lesson Grade: _____     Instructor's Initials: _____

## Assigned Job 32-3
# Practicing Resistance Spot Welding Mild Steel

## Objective:

In this job, you will practice resistance spot welding mild steel.

> **Note**
> Do not attempt this job until you have read all safety precautions, satisfactorily completed the *Resistance Welding Safety Test*, and been approved by your instructor.

1. Your instructor will select the thickness of metal to be welded. The two pieces to be welded do not have to be the same thickness. Obtain about 20 pieces of metal that measure 1″ × 3″ (25 mm × 75 mm). If two different thicknesses are to be welded, obtain about 10 pieces of each thickness. A different material may be specified by your instructor, as well.

2. Clean the material to remove any rust, dirt, or grease. This can be done using steel wool, a wire brush, or a degreasing solution.

3. Before setting up the welding machine, refer to Figure 32-1 in the text and answer the following questions.
   A. What electrode diameter (D) should be used? _____
   B. What tip face diameter (d) should be used? _____
   C. How much force should be used? _____
   D. How many cycles of weld time should be used? _____
   E. How much current should be used? _____

4. Select a pair of electrodes and install them in the machine. Make sure they are installed properly.

5. Adjust the air or hydraulic pressure, or the force spring, to obtain the correct force, determined in step 3C.

6. Set up a welding schedule. Set the squeeze and hold times for 30 cycles each. Set the off time to zero. Set the weld time to the amount calculated in step 3D.

7. If the welding machine is water-cooled, turn on the cooling water.

8. Review the text on setting the proper current. Adjust the tap switch and percent heat controls as required to make good-quality spot welds.

9. Weld together six pairs of metal pieces. Show these pieces to your instructor.

   *Instructor's initials:* _____

10. Tear each pair apart. Are all six welds about the same size? _____

**Caution**
When tearing a weld apart, use a protective screen to protect other workers from flying metal.

11. Write down the tap switch setting, percent heat, and the force you used.
    A. Tap setting: _____
    B. Percent heat: _____
    C. Force: _____

## Inspection:

Show your instructor the pieces after they are torn apart. Your instructor will look at the quality of the welds and give you a grade.

## Assigned Job 32-4
# Making Resistance Spot Welds in Aluminum

## Objective:

In this job, you will resistance spot weld aluminum.

**Note**

Do not attempt this job until you have read all safety precautions, satisfactorily completed the *Resistance Welding Safety Test*, and been approved by your instructor.

1. Your instructor will select the thickness of aluminum to be welded and the appropriate welding conditions. The two pieces do not need to be the same thickness. Obtain about 20 pieces of aluminum that measure 1″ × 3″ (25 mm × 75 mm). If two different thicknesses will be welded, obtain 10 pieces of each thickness.

2. Answer the following questions, using the values for mild steel (even though you are welding aluminum).

    A. What tip diameter should be used? _____

    B. How much force should be used? _____

    C. What weld time should be used? _____

    D. What welding current should be used? _____

3. What welding conditions are provided by your instructor?

    A. What tip diameter should be used? _____

    B. How much force should be used? _____

    C. What weld time should be used? _____

    D. What welding current should be used? _____

4. Compare the values for welding aluminum in step 3 with the values determined for welding steel in step 2. Are the values for welding aluminum greater, the same, or smaller than the values required to weld steel? Circle the appropriate response for each value.

    A. Tip diameter:      Greater     Same     Smaller

    B. Force:      Greater     Same     Smaller

    C. Weld time:      Greater     Same     Smaller

    D. Weld current:      Greater     Same     Smaller

5. Select electrodes with the tip diameters from step 3 and install them into the machine. Make sure they are installed properly.

6. Set the force, weld time, and weld current. Set the squeeze and hold times for 30 cycles each. Set the off time to zero.

7. Turn on the cooling water.

8. Completely clean both surfaces of the aluminum. This can be done using steel wool, a wire brush, or a degreasing solution.

9. Dress the electrodes. After dressing, wipe the electrodes with a cloth to remove any foreign material from the tip faces.

10. Overlap two pieces of aluminum and make a spot weld.

11. Adjust the welding current as necessary to obtain the best quality spot weld. The best welding conditions should be close to the values provided by your instructor. If the aluminum does not weld or a small nugget is created, increase the current. If the weld is too hot and an expulsion weld occurs, reduce the welding current. If a hole is blown in the aluminum, it is probably because there was foreign material on the tip faces.

12. Make additional welds and tear them apart. Make adjustments as required to obtain a good weld nugget diameter.

13. When the machine is set up correctly, weld six pairs of metal pieces. Show these pairs to your instructor.

    *Instructor's initials:* _____

14. Tear each pair apart. Are all six welds about the same size? _____

**Caution**
When tearing a weld apart, use a protective screen to protect other workers from flying metal.

15. Write down the settings that you used.
    A. Tap setting: _____
    B. Percent heat: _____
    C. Weld time: _____
    D. Force: _____

## Inspection:

Show your instructor the pieces after they are torn apart. Your instructor will look at the quality of the welds and give you a grade.

# Section 8
## Welding in Industry

### Lesson 33
## Welding Pipe and Tube

## Objectives:

You will be able to discuss the various processes used to weld pipes and tubes. You will also be able to describe the procedures used to weld pipes and tubes using the SMAW process.

## Instructions:

Read Chapter 33 and study Figures 33-1 through 33-39. Then, answer or complete the following questions.

_____ 1. Which of the following statements is true for tubes?
  A.  Tubes are only available in welded form.
  B.  The wall thickness of a 3″ (75 mm) tube is thinner than the wall thickness of a 3″ (75 mm) pipe.
  C.  Tubes are ordered by their inside diameter.
  D.  Tubes are generally joined by threaded fittings.

2.  *True or False?* When restarting a SMAW weld without a backing ring, the electrode is pushed through the keyhole and held for a moment, and then withdrawn.

3.  What welding process is being used to weld the pipe in the following illustration?

_____

4. What diameter SMAW electrode is suggested for the root pass on a pipe?

_____

5. *True or False?* To properly end a weld, the electrode direction of travel is reversed.

_____    6.    ____ welding on a pipe or tube is done from the 12 o'clock to the 6 o'clock position.
A.  Uphill
B.  Downhill
C.  Freehand
D.  Forehand

7. *True or False?* Pipes are ordered by inside diameter (ID) measurements.

_____    8.    How long is a tack weld made on pipes less than 12″ in diameter?
A.  Less than 1/8″ (3 mm).
B.  1/8″–1/4″ (3 mm–6 mm).
C.  1/4″–1/2″ (6 mm–13 mm).
D.  1/2″–3/4″ (13 mm–20 mm).
E.  3/4″–1″ (20 mm–25 mm).

_____    9.    What device is used to control penetration when welding pipe or tube?
A.  A socket connection.
B.  A pipe collar.
C.  An alignment fixture.
D.  A pipe cutter.
E.  A backing ring.

10. *True or False?* When welding an alloy steel, the passes should be made thinner and wider to reduce the amount of alloy loss.

11. In the spaces provided, put the correct name for each of the six passes of a multiple pass weld.

1st pass: _____

2nd pass: _____

3rd pass: _____

4th pass: _____

5th pass: _____

6th pass: _____

_____    12.    The root reinforcement on the inside of the pipe or tube should be about _____.
A higher root reinforcement would interfere with flow through the pipe.
A.  1/16″ (1.5 mm)
B.  1/8″ (3.0 mm)
C.  3/32″ (2.5 mm)
D.  5/32″ (4.0 mm)
E.  1/4″ (6.0 mm)

13. *True or False?* The pins on a backing ring are generally removed after the joint is tack welded.

14. Tubing may be round, _____, _____, or oval in cross section.

_____

Name _____

_____ 15. What does a large keyhole indicate?
      A. A forward motion that is too fast.
      B. Too much penetration.
      C. Too little penetration.
      D. The need for more current.
      E. The need for a larger electrode.

_____ 16. Which direction of vertical welding usually produces the strongest welds?
      A. Uphill.
      B. Downhill.

17. What action must be taken between SMAW passes to ensure good fusion and a lack of inclusions?

_____

_____

_____ 18. When two fixed, horizontal pipes or tubes are welded, the welding must be done in what position?
      A. Flat.
      B. Uphill or downhill.
      C. Overhead.
      D. All of the above.
      E. None of the above.

19. *True or False?* A 6″ schedule 80 pipe has a greater wall thickness than a 6″ schedule 40 pipe.

_____ 20. A _____ butt joint is used on pipe and thick-walled tubing.
      A. square groove
      B. flair groove
      C. T, K, or Y
      D. V-groove

## Assigned Job 33-1
# Welding a V-Groove Butt Joint on Pipe in the Horizontal Rotated (1G) Position

### Objective:
You will learn to use the SMAW process to make an acceptable V-groove butt weld in the horizontal rotated, or 1G, position.

---

**Note**

Do not attempt this job until you have read all safety precautions, satisfactorily completed the *Shielded Metal Arc Welding (SMAW) Safety Test*, and been approved by your instructor.

---

**Note**

Some instructors will have you practice on an outside corner joint in plate prior to welding on pipe. Pieces of 1/4″ (6.3 mm) or 3/8″ (9.5 mm) plate are used to simulate schedule 40 and schedule 80 pipe.

---

1. You will need five pieces of 4″–6″ (100 mm–150 mm) diameter, schedule 40 or schedule 80, low-carbon steel pipe. The pieces should be 4″ (100 mm) long with a standard wall thickness of .237″–.432″ (6.0 mm–11.0 mm). You will also need four to nine E6010 electrodes and ten to forty E7018 electrodes of the correct diameter. The exact number of electrodes needed depends on the diameter and schedule of pipe.

2. Before proceeding with the exercise, answer the following questions. Refer to Figures 11-6 and 11-9 in the text as necessary.
   A. What diameter and schedule pipe is being used? _____″ (_____ mm), schedule _____.
   B. What is the wall thickness of the pipe? See Figure 33-1. _____″ (_____ mm).
   C. What diameter electrode is suggested for 1/8″–1/4″ (3.2 mm-6.4 mm) thickness? _____″ (_____ mm).
   D. What amperage range is suggested for the chosen electrode diameter?
      E6010:_____ amperes to _____ amperes.
      E7018:_____ amperes to _____ amperes.
   E. Which DC polarity is used with E6010 and E7018 electrodes? _____

3. Set the welding machine amperage for the middle of the suggested amperage range. Set the correct DC polarity.

4. Bevel the ends of the pipe, as required by the drawings in steps 5 and 7.

5. Place two pieces of pipe together to form the weldment shown in the drawing. Set a 3/32″ (2.5 mm) root opening. Tack weld the pipe joint in four places 90° apart. Each tack weld needs to be 1/2″ to 3/4″ (13 mm–19 mm) long. (If practicing on plate, set a 700 included angle. Set a 3/32″ (2.5 mm) root opening.)

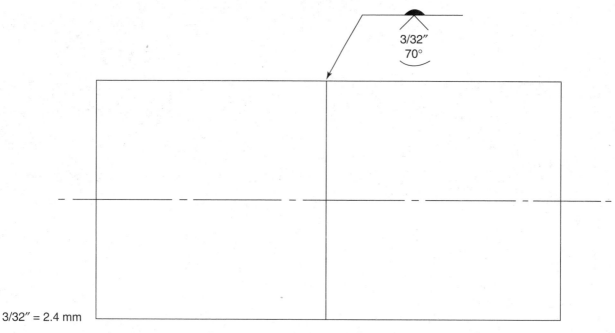

3/32″ = 2.4 mm

6. Clamp the weldment so the pipe is in a horizontal position. Begin the weld at about the 10:30 o'clock position on the pipe. Weld the root pass to the 1:30 o'clock position. Clean the weld thoroughly. Turn the pipe as shown in the drawing and weld the second quarter of the pipe. Clean the weld, then weld the third and fourth quarters as shown in the following illustration.

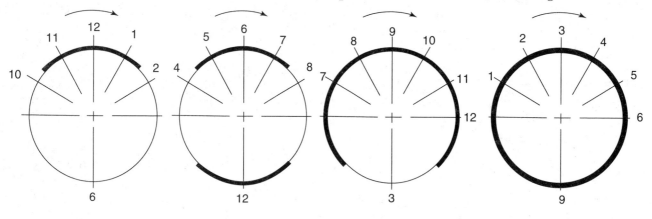

7. After completing the root pass, completely clean the weld. Use E7018 electrodes for the hot and fill passes. Weld from 10:30 to 1:30 for the hot pass. Stop and rotate the pipe 90°. Continue welding the same hot pass again from 10:30 to 1:30. Rotate two more times to complete the hot pass. Completely clean the weld after each stop and after each weld pass. Continue making fill passes until the groove weld is filled and has a 1/16″–1/8″ (1.5 mm–3.0 mm) face reinforcement. All restarts must be made smoothly so the weld bead looks good and has maximum strength.

Name _____

8. Tack weld the other three pieces of pipe to form the weldment shown in the following illustration.

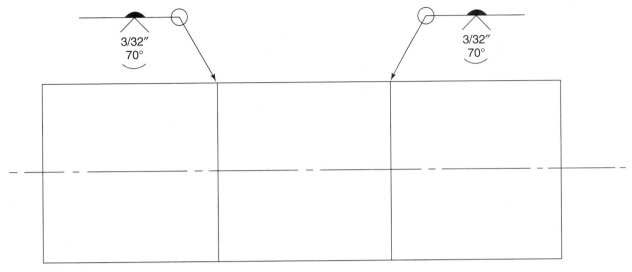

3/32″ = 2.4 mm

9. Weld these joints as in steps 6 and 7.

10. Clean each weld bead with a chipping hammer and a wire brush.

**Caution**
Always wear approved chipping goggles when chipping or wire-brushing the slag from a weld bead.

## Inspection:

All weld beads should be straight and have an even width. The weld beads must be the right size and shape. No undercut or overlap should be present. There should be 100% penetration at the root of each weld. The ripples on the weld bead should be even, bullet-shaped, and have no low spots. Restarts should look the same as the rest of the weld.

## Assigned Job 33-2
# Welding a V-Groove Butt Joint on Pipe in the Horizontal Fixed (5G) Position

### Objective:
You will learn to use the SMAW process to make an acceptable V-groove butt weld in the multiple fixed, or 5G position.

**Note**
Do not attempt this job until you have read all safety precautions, satisfactorily completed the *Shielded Metal Arc Welding (SMAW) Safety Test*, and been approved by your instructor.

**Note**
Some instructors will have you practice on an outside corner joint in plate prior to welding on pipe. Pieces of 1/4" (6.3 mm) or 3/8" (9.5 mm) plate are used to simulate schedule 40 and schedule 80 pipe. The plate can be mounted in various angles for practice.

1. You will need five pieces of 4"–6" (100 mm–150 mm) diameter, schedule 40 or schedule 80, low-carbon steel pipe. The pieces should be 4" (100 mm) long with a standard wall thickness of .237"–.432" (6.0 mm–11.0 mm). You will also need four to nine E6010 electrodes and ten to forty E7018 electrodes of the correct diameter. The exact number of electrodes needed depends on the diameter and schedule of pipe.

2. Before proceeding with the exercise, answer the following questions. Refer to Figures 11-6 and 11-9 in the text as necessary.
   A. What diameter and schedule pipe is being used? _____" (_____ mm), schedule _____.
   B. What is the wall thickness of the pipe? _____" (_____ mm).
   C. What diameter electrode is suggested for 1/8"–1/4" (3.2 mm–6.4 mm) thickness? _____" (_____ mm).
   D. What amperage range is suggested for the chosen electrode diameter?
      E6010:_____ amperes to _____ amperes.
      E7018:_____ amperes to _____ amperes.
   E. Which DC polarity is used with E6010 and E7018 electrodes? _____

3. Set the welding machine amperage for the middle of the suggested amperage range. Set the correct DC polarity.

4. Bevel the ends of the pipe, as required by the drawings in steps 5 and 8.

5. Place two pieces of pipe together to form the weldment shown in the following drawing. Set a 3/32″ (2.5 mm) root opening. Tack weld the pipe joint in four places 90° apart. Each tack weld needs to be 1/2″ to 3/4″ (13 mm–19 mm) long. If practicing on plate, set a 70° included angle. Set a 3/32″ (2.5 mm) root opening.

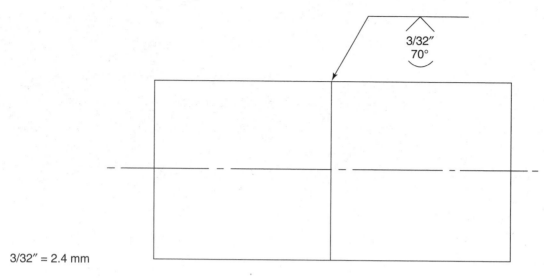

3/32″ = 2.4 mm

6. Place the weldment in a weld positioner so that the centerline of the pipe is horizontal. This weld must be made all around the joint with the pipe in a fixed position.

7. Begin welding the root pass on one side of the pipe at the 6 o'clock position and weld to the 12 o'clock position. Thoroughly clean the weld. Then begin again on the other side of the pipe at the 6 o'clock position and weld to the 12 o'clock position to finish the root pass. See the following drawing.

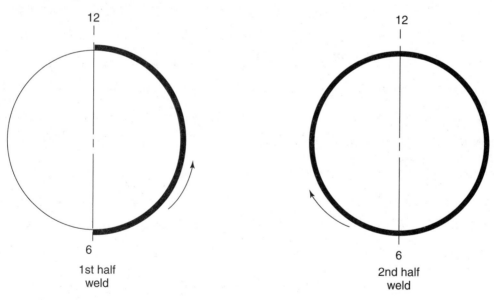

1st half weld        2nd half weld

8. After completing the root pass, completely clean the weld. Use E7018 electrodes for the hot and fill passes. Repeat the process of welding from 6:00 to 12:00 on each side of the pipe for additional fill passes. Completely clean the weld after each stop and after each completed fill pass. Continue making additional fill passes until the groove weld is complete and there is 1/16″–1/8″ (1.5 mm–3.0 mm) reinforcement. All restarts must be made smoothly so the weld bead looks good and has maximum strength.

Name _____

9.  Tack weld the other three pieces of pipe to form the weldment shown in the following drawing.

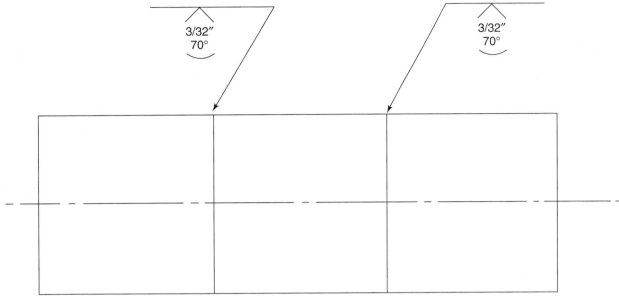

3/32″ = 2.4 mm

10.  Weld these joints as described in steps 7 and 8.

11.  Clean the weld beads with a chipping hammer and a wire brush.

**Caution**
Always wear approved chipping goggles when chipping or wire-brushing the slag from a bead.

## Inspection:

All weld beads should be straight and have an even width. The weld beads must be the right size and shape. No undercut or overlap should be present. There should be 100% penetration at the root of each weld. The ripples on the weld bead should be even, bullet-shaped, and have no low spots.

# Lesson 34
# Robotics in Welding

## Objectives:

You will be able to describe the types of robots used in the welding industry. You will be able to name the parts and identify the welding axes of a welding robot. You will also be able to discuss the safety precautions to take when working around a robot.

## Instructions:

Read Chapter 34 and study Figures 34-1 through 34-9. Then, answer or complete the following questions.

1. Define the term *actuator*.

   _____

   _____

2. List four (GMAW) welding parameters that the welding programmer or operator must set into the computer memory.

   _____

   _____

3. On an electrically actuated robot, what type of electric motors are used to move and stop the actuators at very precise distances?

   _____

4. *True or False?* Once the robot program is started, the program *cannot* be interrupted.

5. A robot _____ checks and monitors itself as well as the welding operation.

   _____

6. *True or False?* Operators and workers must never enter the working volume of a welding robot while the robot is energized.

7. List three reasons for using robots for welding.

   _____

   _____

   _____

   _____

8. *True or False?* Some electric robots can repeat a movement with an accuracy of ±.004″ (.10 mm).

9. Name the parts of the robotic welding workstation shown in the drawing.

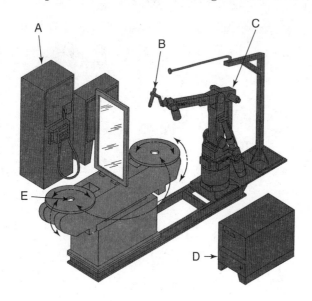

A. _____

B. _____

C. _____

D. _____

E. _____

10. *True or False?* The robot must often reach out toward the welding area. Most of this reaching motion is done by rotating the shoulder axis.

_____ 11. Robotic systems can be used with which of the following processes?
   A. Plasma arc cutting.
   B. Resistance spot welding.
   C. Laser beam welding and cutting.
   D. Gas tungsten arc welding.
   E. All of the above.

12. *True or False?* A parts positioner is often used to move parts into and out of the robot cell.

_____ 13. What is the term for the item shown here?
   A. The robot controller.
   B. The positioner.
   C. The teach pendant.
   D. The robotic command unit.

Name _____

14. *True or False?* The main purpose of a positioner is to make the robot more efficient.

15. The operator must use the ____ ____ to move the robot and welding torch through all of the movements required to complete the welds.

   _____

16. *True or False?* The robot working volume is the three-dimensional space occupied by the robot at the extremes of its movements.

17. List three of the five factors that must be considered when choosing a welding robot for a particular job.

   _____

   _____

   _____

   _____

18. *True or False?* The linear (straight-line) or horizontal reaching motion of a robot is normally accomplished by rotating the robot around its waist joint.

19. A robot cell contains all the parts of the welding robot system. Describe two ways to protect workers or prevent them from entering a cell while a robot is working.

   _____

   _____

   _____

20. Name the axes shown in the following drawing of a six-axis robot.

   A._____

   B._____

   C._____

   D._____

   E._____

   F._____

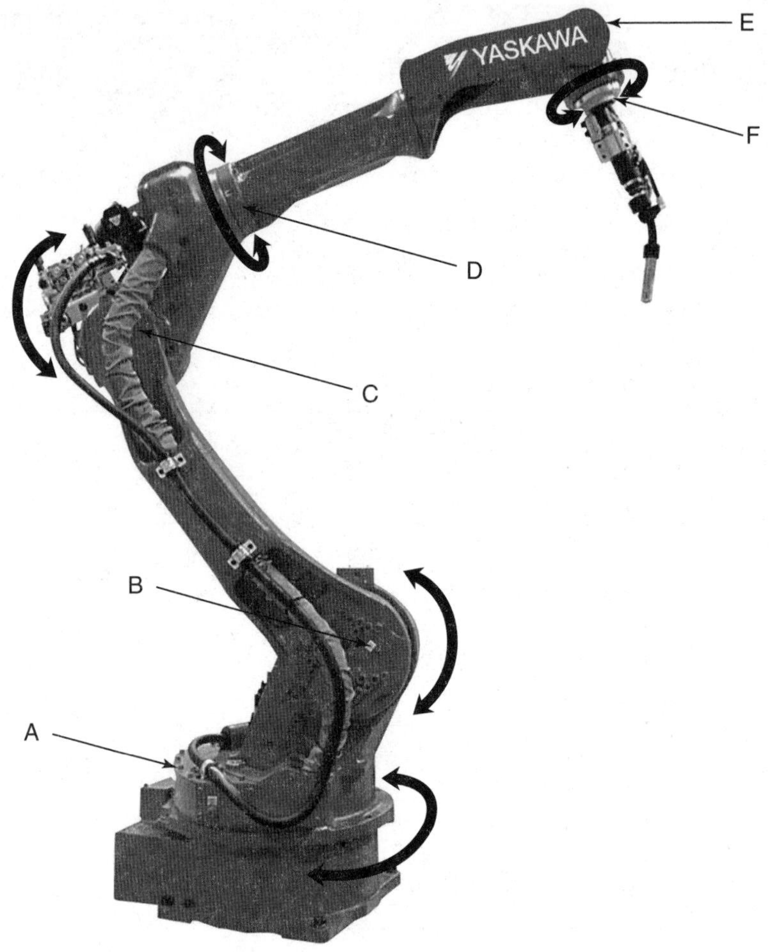

Name: _____    Date: _____

Class: _____    Instructor: _____

Lesson Grade: _____    Instructor's Initials: _____

# Lesson 35
# Special Welding and Cutting Processes

## Objectives:

You will be able to describe the principles of operation of several special welding and cutting processes and how and when these special processes are applied. You will also be able to identify several welding processes by their AWS abbreviations.

## Instructions:

Read Chapter 35 and study Figures 35-1 through 35-25. Then, answer or complete the following questions.

_____ 1. What is the maximum metal thickness that can be welded using the EGW process?
   A. 9″ (229 mm).
   B. 12″ (305 mm).
   C. 14″ (356 mm).
   D. 18″ (457 mm).
   E. Virtually any thickness.

_____ 2. Metals may be heated to the welding temperature by rapidly vibrating them. What is this welding process called?
   A. FRW.
   B. USW.
   C. EXW.
   D. LBW.
   E. EBW.

3. The _____ protects the molten metal from oxidation in the ESW process.

_____

4. *True or False?* Hollow electrodes and compressed air are used to cut underwater using the OAC process.

5. Metals can be welded using a high energy-density beam of light. What is the AWS abbreviation for this process?

_____

6. *True or False?* Shielding gas can be used with a stud welding gun to produce good-quality stud welds on nonferrous metals.

7. What is the name and abbreviation for the cutting process that uses an arc between a carbon electrode and a base metal with high-pressure air from the electrode holder blowing molten metal out of the cutting area?

_____

8. What part of the EGW or ESW equipment prevents molten metal from running out of the weld area?

_____

9. What is the highest welding temperature reached in the PAW processes?

_____

10. The fuel rods in the electrodes used in exothermic cutting come in small diameters, called _____, and in larger diameters, called _____.

_____

11. *True or False?* An oxygen jet is sometimes used with laser beam equipment to drill or pierce.

_____ 12.  Which of the following is not an advantage of laser beam and electron beam welding processes?
    A.  Low heat input.
    B.  The ability to weld dissimilar metals.
    C.  The ability to weld parts as thin as .001″–.002″ (.025 mm–.050 mm).
    D.  The ability to achieve complete penetration in metals over 12″ (305 mm) thick.

_____ 13.  What cutting process can be used to cut through concrete?
    A.  Oxygen arc cutting.
    B.  Exothermic cutting.
    C.  Air carbon arc cutting.
    D.  None of the above.

_____ 14.  Which of the following is true for the transferred arc PAW process?
    A.  In the transferred arc process, the filler metal is transferred to the base metal from the electrode.
    B.  In the transferred arc process, the arc is transferred from the electrode to the torch nozzle.
    C.  In the transferred arc process, the arc is transferred from the electrode to the base metal.
    D.  In the transferred arc process, the filler metal is transferred to the torch nozzle.
    E.  None of the above.

15. One welding process uses a rotating tool to heat and mix base metals. It is capable of full penetration welds. What is the name of this process and its AWS abbreviation?

_____

16. *True or False?* SMAW *cannot* be done underwater.

17. *True or False?* An ultrasonic weld is completed faster than a resistance spot weld.

18. Name three inert gases that may be used to create the plasma in the PAW process.

_____

_____

19. *True or False?* Hot air is used to weld thermoplastics. There is no flame or arc.

_____ 20.  Which of the welding processes listed below uses an arc that is under a thick layer of flux?
    A.  Electrogas welding (EGW).
    B.  Submerged arc welding (SAW).
    C.  Plasma arc welding (PAW).
    D.  Friction stir welding (FSW).

Name: _____     Date: _____

Class: _____     Instructor: _____

Lesson Grade: _____     Instructor's Initials: _____

## Lesson 36
# Inspecting and Testing Welds

## Objectives:

You will be able to determine a weld flaw and a weld defect. You will be able to prepare weld samples for testing. You will be able to use shop methods to inspect and test sample welds. You will also be able to list common destructive and nondestructive testing methods.

## Instructions:

Read Chapter 36 and study Figures 36-1 through 36-25. Then, answer or complete the following questions.

1. *True or False?* Any flaw in a weld will require the weld to be removed and rewelded.

2. *True or False?* Any defect in a weld makes the weld unusable.

3. *True or False?* A weld with a defect must be removed and rewelded.

4. *True or False?* Tests may be divided into two groups, destructive and nondestructive.

5. *True or False?* The most common of all nondestructive tests is X-ray inspection.

6. *True or False?* The abbreviations NDE and NDT can be used interchangeably.

7. *True or False?* Surface flaws may be detected using dye-penetrant inspection.

8. *True or False?* Flaws that are deep within the weld *cannot* be found using magnetic-particle inspection.

9. *True or False?* The ultrasonic inspection process uses sound waves to locate defects within a weld.

10. *True or False?* The fillet test is a nondestructive testing method used on fillet welds.

11. A(n) ____ is a flaw that makes the weld unusable for its intended use.

   _____

12. The welding ____ defines how large a flaw may be before it is considered a defect.

   _____

13. When preparing a test sample of a plate or pipe section for a tensile test, ____ inches is the specified width of the reduced section.

   _____

14. While tensile testing a weld sample measuring .75″ (19.0 mm) in diameter, the required load to break it is determined to be 13,000 pounds (58,000N). What is the calculated tensile strength of this sample?

_____

_____

15. The test sample from the previous question is marked 1″ (25 mm) on each side of the center of the sample. After the sample is pulled apart, the pieces are placed together and measured again. The second measurement is 1.240″ (31 mm). What is the calculated ductility (percentages) of the sample?

_____

_____

16. For a guided bend test, the test sample has the root of the weld running across the width of the outer face of the bend. This type of bend test is called a(n) ____ ____ ____.

_____

_____    17.  When grinding the surfaces of a weld sample for a guided bend test, you should ____.
      A.  grind parallel to the direction of the weld
      B.  grind across the weld
      C.  grind the weld in a cross-hatch pattern
      D.  grind the weld in a circular pattern

18. When performing a weld for a fillet weld test sample, at what point along the joint should the weld be stopped and restarted?

_____

19. Describe a peel test on a spot welding sample.

_____

_____

_____

20. List two types of hardness tests.

_____

_____

Name: _____ Date: _____

Class: _____ Instructor: _____

Lesson Grade: _____ Instructor's Initials: _____

## Lesson 37
# Welder Certification

## Objectives:

You will be able to understand the use of welding codes and specifications. You will be able to locate required information in welding codes and specifications. You will be able to explain the differences between welding performance qualification and welding procedure specifications.

## Instructions:

Read Chapter 37 and study Figures 37-1 through 37-6 and the Appendices A, B, and C. Then, answer or complete the following questions.

1. *True or False?* A contract is an agreement between two people or companies.

2. *True or False?* A contract contains only a description of what is to be built, when it is to be completed, and how much it will cost.

3. *True or False?* Codes and specifications cover information on base metals, safety, filler metals, welding variables, and other details.

4. *True or False?* The welding procedure specification (WPS) is used to qualify the welders on the job.

5. *True or False?* Each major change that may affect the quality of a weld must have its own welding procedure specification written.

6. *True or False?* A procedure qualification record is required for each welding procedure specification.

7. *True or False?* Welders often must pass a number of welding performance qualifications. A different test is required to qualify a welder for each electrode type or filler metal being used.

8. *True or False?* A welder who passes a qualification test on one thickness of metal is qualified to weld any thickness of the same metal.

9. *True or False?* Test plates are welded according to the welding procedure specification.

10. *True or False?* Welders are tested periodically to ensure that they are still able to perform the required welds to the required quality.

11. The welding procedure specification is approved when the ____ ____ ____ shows that the WPS results in good welds.

_____

12. A welder who passes a qualification test in the horizontal position is qualified to perform that weld in the _____ and _____ positions.

_____

13. A groove weld made in the overhead position is referred to in ASME code as the _____ position.

_____

14. A fillet weld made in the horizontal position is referred to in ASME code as the _____ position.

_____

15. A groove weld made by rotating the pipe in the flat position is referred to in ASME code as the _____ position.

_____

16. A fillet weld made on pipe welded in the overhead position is referred to in ASME code as the _____ position.

_____

17. How many tension test and guided bend test coupons will be tested to qualify a welder being tested on 1 1/2″ base metal? See Figure 37-3.

_____

18. To qualify as a welder on a given job, the welder must pass the _____ _____ _____ test.

_____

19. The mode of metal transfer (welding method) is specified on what document?

_____

20. After test samples are tested using the WPS, where are the results of these tests recorded?

_____